W9-BMB-990

Making Waves

ALSO BY ROGER LEWIN

Complexity: Life at the Edge of Chaos

The Sixth Extinction (with Richard Leakey)

The Thread of Life

Weaving Complexity and Business (with Birute Regine)

Patterns in Evolution

Making Waves

IRVING DARDIK

and His

SUPERWAVE PRINCIPLE

ROGER LEWIN

This book is intended as a reference volume only, not as a medical manual. The information given here is designed to help you make informed decisions about your health. It is not intended as a substitute for any treatment that may have been prescribed by your doctor or therapist. If you suspect that you have a medical problem, we urge you to seek competent medical help.

Printed in the United States of America

Rodale Inc. makes every effort to use acid-free ♾, recycled paper ♲.

Book design by Christopher Rhoads

Cover photograph © by G.K. and Vikki Hart/Photonica

Library of Congress Cataloging-in-Publication Data

Lewin, Roger.
Making waves : Irving Dardik and his superwave principle / Roger Lewin.
p. cm.
Includes index.
ISBN-13 978–1–59486–044–7 hardcover
ISBN-10 1–59486–044–0 hardcover
1. Dardik, Irving. 2. Scientists—United States—Biography. 3. Surgeons—United States—Biography. 4. Waves. 5. Quantum mechanics. I. Title.
Q143.D27L49 2005
509'.2—dc22 2005018871

Distributed to the trade by Holtzbrinck Publishers

2 4 6 8 10 9 7 5 3 1 hardcover

We inspire and enable people to improve their lives and the world around them

For more of our products visit **rodalestore.com** or call 800-848-4735

"For deriving all from nothing there suffices a single principle."

—GOTTFRIED WILHELM LEIBNIZ

Contents

Foreword

by Michael McKubre, Ph.D., director, Energy Research Center, SRI International, Menlo Park, CA

In this centennial year of Einstein's Special Theory of Relativity, it is appropriate to recall that new ideas are not only possible, they are inevitable and vital to the well-being of science. The leading edge of applied physics, for example, is strong, vibrant, and expanding. At its core, however, our understanding remains at best incomplete, as illustrated by Sir Arthur Eddington's reference to Heisenberg's Uncertainty Principle: "Something unknown is doing we don't know what." It is possible that the ideas contained in the following pages will help us understand if not what the unknown is made from, then at least what it is doing to create everything we know in living and nonliving systems. This would be huge progress toward shoring up our core knowledge of physics.

Few who understand the words on these pages will emerge at the end with their worldview unchanged. This idea is well expressed and not difficult to grasp, and it is not mystical or mathematical. Irving Dardik's simple idea speaks to the coherence of living systems, of physics and physical science. Imagine yourself on the ocean shore. It is easy to appreciate the periodicity of the waves, but waves arrive from many directions. The dashed waves surge out, creating counterwaves. Wave peaks build and subside; waves form within waves. It appears complex but the structure is not chaotic or arbitrary. The pattern of waves is continuously and coherently connected with every other point and influence of, on, and within the ocean. In Dardik's SuperWave Principle, the ocean is the universe. Through the medium of waves, every point communicates with every other, as one body. Chaos and disorder are not invoked; statistics and other mathematical abstractions are not involved.

What does this principle have to do with the nature of life, metallurgy, physics, physical chemistry, the reversal of disease, wellness, athletic performance? More broadly, what promises relief from the deadening drudgery of entropy, inevitable decay, and disorder? Dardik's concepts of waves shed new light on all of these issues and more. This idea could turn out to be the most important new understanding of nature in our age. Roger Lewin has patiently reconstructed the history of Dardik's Big Idea, and has clearly set out examples of real applications, current and pending. The story is so engaging and compelling it might seem fiction, but it is fact. The central characters here, Dardik, Godfrey, Kelly, Kimmel seem larger than life simply because, in life, they are and were that way.

It is true that physical observations or novel concepts that seriously challenge our preconceptions are likely at first to be ignored or marginalized by the scientific establishment. Novelist H. G. Wells recognized sufficient reason for this seemingly unscientific behavior: "New and stirring things are belittled because if they are not belittled, the humiliating question arises, 'Why then are you not taking part in them?'" And yet, new ideas do enter science. Our next generation will simply accept cold fusion as reality. Will they also accept Irving Dardik's SuperWave theory of everything? Forced to guess, I would say yes. Already I accept the evidence, implications, and practical application of the physiological aspects of the principle. Tests of the broader context are still ongoing in many reputable institutions around the world, including my own. But who am I to judge or make such predictions? Read on and draw your own conclusions.

Preface

I have to admit that when I first heard Irv Dardik gush about his SuperWave Principle—and he does *gush* about it—I was skeptical. My four decades as a science writer taught me to be attracted to new ideas, but also to be wary of them, because they can be seductive but ultimately unfounded, particularly when such claims are promoted with exuberance of the kind that Dardik displays. Surely, you might say, if an idea is valid it will stand on its own merits. But my experience in observing many areas of science has taught me that standing on its own merits is often tough for any new idea, especially a *big* new idea.

Challenging tradition is not for the faint hearted. Defenders of the status quo can be fierce in their treatment of perceived upstarts touting what are viewed as ill-considered and even dangerous ideas. I'll give an example in a field of scientific endeavor close to my own interests, human evolution.

By the late 1960s and early 1970s there was a clear and solid consensus among anthropologists that the fateful split between humans and apes had occurred in ancient times, much earlier than 10 million years ago. And they had the fossil evidence to prove it, they claimed: the jaws and teeth of a creature named *Ramapithecus* that lived in Asia some 13 million years ago. Despite being apelike in many ways, the jawbones of this creature were said to display unmistakable anatomical features that clearly screamed "early human." These features were enumerated in great detail in many earnest papers. Case closed. Or so it was thought.

Beginning in 1967, two biochemists from the University of California at Berkeley, Vincent Sarich and Alan Wilson, stepped into the arena and effectively said, "You anthropologists have got it badly wrong: the human/ape split occurred much more recently than 10 million years ago." The response of the anthropological establishment was at first to ignore it, and then when Sarich and Wilson wouldn't shut up, to ridicule it. A big problem for Sarich and Wilson was not only that they were proposing the "wrong" date, but also that they were using the "wrong" evidence. The Berkeley duo weren't fossil experts; they were biochemists who compared certain properties of proteins from humans and living apes and came up with an answer for when we last shared a common ancestor: 5 million, not 13 million, years ago.

One prominent anthropologist of the day fulminated at a scientific meeting about Sarich and Wilson's temerity: "I object to careless and thoughtless statements about evolutionary processes in some of the conclusions drawn from [biochemical] data. . . . Unfortunately there is a growing tendency, which I would like to suppress if possible, to view the [biochemical] approach to human evolutionary studies as a kind of instant [answer]. No hard work, no tough intellectual arguments. No fuss, no muss, no dishpan hands. Just throw some proteins into the laboratory apparatus, shake them up, and bingo!—we have the answer to questions that have puzzled us for at least three generations." Much worse was said in private, as I heard time and again.

Unfortunately for the above, unnamed gentleman, Sarich and Wilson turned out to be right. But it took a few years before all anthropologists could bring themselves to accept the 5 million year date. And it took a few more years before Sarich and Wilson's crime of bringing into the anthropological arena a relatively simple and foreign line of evidence, biochemistry, was seen not to be a crime at all; instead it was to be the way of the future. Molecular anthropology is now a mainstay of the field, with many journals devoted to it.

Sarich and Wilson's idea was right, but it wasn't allowed to stand on its merits because it was too threatening to the establishment. Eventually it prevailed, in no small measure due to Sarich's enthusiasm, super-sized physical presence, forceful personality, and persistence, all of which Dardik has too.

The moral of the story is simple: Just because the establishment dismisses a Big New Idea doesn't mean it is necessarily wrong. (It's also the case, of course, that rejection doesn't guarantee an idea is correct!) The only honest way to proceed is to give the idea a fair hearing; look at what, in practical terms, it achieves; and then decide.

My own evolution with the SuperWave Principle was first to see it as a little wacky sounding; then to think, "Hmm, there might be something in this;" then to seriously consider that this *is* a big idea. Obviously, unless I had made that final step I would not have devoted myself to this book project.

Dardik's Big Idea has many dimensions. The health dimension is the one that's easiest to experience; it just seems right, intuitively. The kind of exercise that Dardik espouses—intermittent—is natural: it is the way that animals typically move, and children, too. The health crisis that looms over Western civilization, in the surge of lifestyle diseases and modern medicine's sclerotic approach to them, requires a revolutionary new approach, and the SuperWave Principle may be just what is required.

But, just as Sarich and Wilson faced heavy skepticism from mainstream anthropologists when they brought in an unconventional technique to solve an age-old question, Irv Dardik frequently finds himself talking to medical folk who listen politely—and sometimes not so politely—to his arguments, and then turn away dismissively. Indeed, Dardik's whole life, particularly in his role as enthusiastic promulgator of the SuperWave Principle, has had many twists and turns, ups and downs, financially and in terms of recognition of his ideas. It has been a very wavelike journey in its own way and an important part of that journey has been the strong financial and moral support of Sidney Kimmel, a New York industrialist.

Kimmel recognized something important in Dardik's venture, and, visionary that he is, stood by Dardik when his advisors and scientific acquaintances told him he was wrong-headed to do so. Kimmel's behind-the-scenes financial support of Dardik's scientific venture, and of other scientific research, particularly cancer research, invites comparison with the imagination and foresight of Alfred Lee Loomis, a little more than six decades ago. Loomis, a Wall Street tycoon, saw the pressing need for clandestine research in support of the war effort against Nazi Germany, which he bank-rolled and helped guide in a laboratory in his New York mansion, an extraordinary effort that was chronicled in the 2002 book *Tuxedo Park*. An important difference between the two men, however, is that while Loomis was constrained to pursue elements of war, Kimmel has grasped the opportunity to use his fortune to enhance peace and health..

Dardik argues that, ultimately, waves are ubiquitous in nature, at all scales, from the universe itself to the smallest elements of matter. It is here, at the lowliest levels of nature, that the principle is toughest to wrestle with, particularly for nonphysicists. For it is here that we enter the quantum world, a world so strange that it boggles the minds of even those who claim to understand it, claim to believe it. Most of us are familiar with the simple model of matter as being made of collections and conglomerations of particles, like ping-pong balls whizzing around in space. But the quantum view of matter is that all is not what it appears: particles may be particles under some circumstances but are waves under others. According to the SuperWave Principle, however, particles are simply manifestations of waves, that *everything* is waves exerting themselves in different ways. "Matter is waves waving," says Dardik. Taoist philosophy comes to mind: the harmonious power of opposing forces, two sides of a wave, which restores balance and order in life.

This book tells Irving Dardik's personal story and the evolution of his SuperWave Principle, which leads into some surprising byways of science. For example, the Principle is finding practical application in achieving more efficient stirring of molten metals, a hugely frustrating and expensive preoccupation in the world of metallurgy. It is also showing promise in the promotion of low-energy nuclear reactions, or, as it is colloquially called, cold fusion.

"Wait a minute," you might say. "Cold fusion? Wasn't that debunked a decade and a half ago? Wasn't it shown to be a fraudulent illusion of unhinged minds, the product of pathological science?" Good questions. I thought the same when I embarked on this project. But when I probed into the field I discovered a vigorous research community producing impressive results in persevering with the goal of unleashing limitless energy, a pressing concern in today's world. While cold fusion research is regarded as a legitimate scientific pursuit in Europe and Japan, in the United States it is nothing less than an academic pariah. However, the U.S. Department of Energy recently completed a favorable review of the field, recommending renewed funding. Cold fusion is far from dead; it is thriving out of sight, perhaps soon to reemerge as real science.

By applying the SuperWave Principle to cold fusion, physicists in Israel under Dardik's inspiration and supervision have produced many times more energy in single experiments during the past three years than anyone else in the world has achieved in similar ventures in the past fifteen years. If the SuperWave Principle is as powerful as Dardik claims, this kind of success is just the beginning of a long line of exciting breakthroughs in the field of human health, as well as in many different areas of science and technology.

As children and as adults we are told "Don't make waves!" In other words, "Be good, don't upset the prevailing order of things." We know, of course, that in order to make progress in science, you have to make waves, as philosopher Thomas Kuhn said in his *Structure of Scientific Revolutions*. And yet too often the reward for making waves in science is derision from peers, as pioneers through the ages have experienced. Think of what happened to Copernicus and Galileo.

Irving Dardik has been making waves for years, literally and figuratively. Literally, in his novel exercise program, which make waves in people's physiology, causing health. And figuratively, by challenging received notions about the meaning of health and the nature of the universe. As a result, he's experienced his share of criticism, particularly in a clash with the medical system. And he has experienced resistance from physicists to his radical notion of how the universe works. But Irving Dardik is nothing if not persistent. Today, that persistence is beginning to pay off.

PART ONE
The Nature of Nature

1

It's All Waves

"Maybe there are no particle positions and velocities, but only waves. It is just that we try to fit the waves to our preconceived ideas of positions and velocities."

—STEPHEN HAWKING

EVER SINCE LIFE FIRST APPEARED ON EARTH, some 3 billion years ago, the sun has risen and set a trillion times, a constant daily rhythm to which the vast majority of organisms are exposed. It is little wonder, then, that virtually every aspect of an individual organism's behavior and physiology is imprinted in some fashion by daily, or circadian, rhythms. Scientists have been fascinated with these rhythms for centuries, but for much of the 1800s and 1900s the scientific establishment balked at the notion. It is only in the past few decades that such rhythms have been universally acknowledged to exist. The principal reason for the resistance to the notion of rhythmic fluctuations in human physiology was the belief that normality in nature was characterized by stability and that deviations from stability implied a state of pathology. Scientists now know that these rhythms are real and not simply driven by the cycle of dark and light; they are actually are encoded in our genes—an internal biological clock.

The French astronomer and mathematician Jean Jacques Dortous de Mairan was the first scientist to perform an experiment on circadian rhythms, in the early 1720s. He noticed that the mimosa plant by the window in his Paris study stretched its leaves toward the sun in the morning and folded them up when darkness fell. Wondering whether the plant was simply responding to the sun's rays, de Mairan picked up the plant and put it in a dark closet. The next morning he looked into the closet and saw that the plant's leaves were unfurled, just as they had been when it was located by the window. When the sun went down he looked again;, the plant was acting just as it had while in the open: the leaves were folded up. He repeated the experiment with other "sun-sensitive plants" and found that they, too, reacted in the same way in the dark closet: they unfurled their leaves in the morning and folded them as the sun set.

De Mairan, whom Voltaire described as one of the five most outstanding scientists of the 18th century, didn't write up his results, claiming he had more important scientific work to occupy his time. But his botanist friend Jean Marchant, to whom de Mairan had described his observations, decided that the finding was so important that *he* would tell the world about it at a scientific meeting in Paris in 1729. He described the mimosa plant's sensitivity to light and dark, explaining how it unfurled and folded its leaves in synchrony with the sun's rise and fall. "But," he went on, "Monsieur de Mairan observes that this reaction can be observed even if the plant is not in the sunlight or outdoors."

Two millennia earlier, the Greek scientist Androsthenes, who chronicled Alexander the Great's expedition to India, had similarly noticed a plant's response to day and night. He wrote that by raising its leaves in the morning, the tamarind tree appeared to be "worshipping the sun." Romantically appealing though this notion might have been, de Mairan's scientific experiment showed it to be wrong because plants raise their leaves even when no sun is visible. Marchant observed that the plant "feels the sun without in any way seeing it." That, too, was wrong; had de Mairan persisted with his observations over a period of weeks, rather than a few days, he would have found that although the plant would continue to unfold and fold its leaves in a regular way, the cycle would slip to a little less than 24 hours. (Such an experiment wouldn't be done for another century.) What this means is that the plant has its own *internal* clock that it follows slavishly in the absence of natural light and dark, on a 23-hour cycle. The effect of exposure to the 24-hour day/night cycle is to *entrain* the internal clock to run on a 24-hour schedule, in synchrony with the rising and setting sun.

In his presentation to fellow scientists in Paris, Marchant extended his observations to humans. He said that the mimosa plant's behavior "seems to be related to the delicate sensitivity by which invalids confined to their sickbeds perceive the difference between day and night." Humans, like light-sensitive plants, know when it is night and when it is day, even when they are isolated from natural light, he said. Brilliant as this insight was, it was also wrong in the same way his statement that the mimosa plant "feels the sun without in any way seeing it" was wrong. Like mimosa plants, humans possess an internal clock that is slightly different than 24 hours—ours is a little longer—but this wasn't demonstrated until the middle of the 20th century. In numerous experiments, volunteers spent weeks in isolation—in bunkers, in caves, cut off from natural light, without social interaction from the outside, and without access to clocks. They recorded when they slept and when they woke. In all cases, their records showed that the "natural" daily rhythm for humans is close to 25 hours. Like the mimosa plant, a human's internal clock is entrained to a 24-hour cycle when exposed to natural light and dark.

From the 1960s on, scientific interest in circadian rhythms burgeoned, revealing daily fluctuations in practically every behavior and physiological measure that was investigated. The activity/sleep cycle is obvious, of course. But our physical and mental activities fluctuate, too. Hand-eye coordination is best in the early afternoon; mental alertness and athletic ability are highest in the mid to late afternoon—which, incidentally, is when many world records are set.

In the physiological realm, we think of "normal" body temperature as being 98.6°F: it's inscribed on the clinical thermometer in your bathroom cabinet. Actually, every healthy individual's temperature varies throughout the day, from as low as 96° to as high as 100°F. Body temperature is lowest a couple of hours before your regular waking time; by the time you wake, it has already begun to rise. It plateaus for a while in midafternoon, reaches a maximum around 7:00 P.M., and then starts to fall.

Similarly, blood pressure and pulse rate fluctuate with the time of day. The production and release of scores of hormones and enzymes follow the daily cycle, but with each element marching to its own beat. The human body is like a symphony orchestra, with each instrument (physiological function) following its own characteristics but contributing to the harmony of the whole.

Some disease symptoms also follow a daily clock, as the Greek poet Hesiod noted in 700 BCE: "Of themselves, diseases come among men, some by day, some by night." We now know that heart attacks, strokes, headaches, hay fever, and rheumatoid arthritis cluster in the morning. Symptoms of asthma, gout, colic in infants, gastric ulcers, and heartburn tend to occur at night.

The study of all these fluctuations—our circadian rhythms—is known as chronobiology, and it is a vigorous field of scientific endeavor. Scientists are not only trying to understand how the rhythms work, but are also exploring the health impact of the disturbance of rhythms. They are seeking ways to enhance medical treatment by, for instance, administering drugs at a time of day when they are most effective, an approach known as chronotherapy.

Prior to the 1950s, biologists were ignorant of all these daily rhythms and their implications. There wasn't even a word for them! The term *circadian*—from the Latin *circa*, meaning "about," and *diem*, meaning "day"—was coined in 1959 by the American biologist Franz Halberg, who is regarded as the father of chronobiology. Biologists had overlooked circadian changes in physiological measures for the simple reason that, according to the prevailing paradigm, such changes were not supposed to happen. Toward the end of the 19th century, the French physiologist Claude Bernard argued that the body is built so that it tries to maintain constancy in its *milieu interieur*, or internal environment. Half a century later, in the late 1920s, the Harvard physiologist Walter Cannon enshrined Bernard's idea as the principle of *homeostasis*, or steady state.

The body's physiology, Cannon argued, is finely tuned to respond to any deviation from constancy, by bringing the factor in question back to "normal." For the better part of the 20th century, homeostasis dominated teaching in biology and medicine, and still does in many ways. It is true, of course, that the human body, like all organisms, has mechanisms that prevent physiological systems from slipping dangerously out of kilter, which Cannon called "the wisdom of the body." But the principle of homeostasis was taken too far, to include the notion that *any* fluctuation in a physiological measure was wayward. Moreover, the hypnotic quality of the *idea* of homeostasis encouraged biologists to believe that *stability* was a sign of normality and health; any departure from stability, such as regular or irregular rhythms, was taken as a sign of abnormality or ill health. Now, with four decades of research on circadian rhythms recorded in the scientific literature, it is clear that the opposite is true: rhythmic fluctuation is ubiquitous and a sign of normality and health; the absence of fluctuation betrays abnormality and ill health.

Recognizing the existence of circadian rhythms was an important step toward understanding how nature works. Finding out what propels those rhythms came next. A natural and reasonable assumption was that exposure to day/night cycles was the ultimate engine of circadian rhythms. That's what de Mairan and Marchant concluded back in the 18th century—and it's what the great majority of scientists believed, until relatively recently. Three decades ago, scientists discovered what may be described as the conductor of the human body's orchestra of daily rhythms: two groups of about 10,000 nerve cells, called the suprachiasmatic nucleus (SCN), tucked away at the base of the brain in the bean-size hypothalamus. The hypothalamus, sometimes called the body's master gland, helps regulate breathing, heart rate, body temperature, hormone production, and many other important physiological functions. The role of the SCN is to impose a circadian rhythm on all the physiological functions under the control of the hypothalamus. When the SCN is destroyed—in a laboratory experiment, for example—the animal's daily rhythms vanish.

The steady beat of the circadian rhythm is built into the very nature of the SCN—an inner, master clock, so to speak, that runs on a cycle slightly longer than 24 hours, independent of the day/night cycle. But, as mentioned earlier, the SCN's cycle is *entrained* by light. Certain light-sensitive cells in the retina (different from the ones that mediate vision) send signals to the SCN. As a result, our daily cycle follows the external

24-hour rhythm. If entrainment were not possible, then we wouldn't be able to adjust when we travel to a different time zone, and jet lag would be permanent.

During the 1990s, molecular biologists discovered a handful of genes that underlie circadian rhythm of the SCN. And these same genes operate in all organisms, including fruit flies, scientists's favorite organism for studying genes and their effects. Much to the surprise of everyone, these same discoveries dislodged the SCN from its supposed all-powerful role as the inner clock. It turns out that the circadian genes are active in all tissues, implying that there are many clocks throughout the body, all following the same rhythm. Place human tissue, for example, in a petri dish, and the cells keep up their circadian rhythm of gene activity, hormone secretion, and energy production. Our bodies, it seems, are suffused with an innate rhythm, pulsating unendingly in every cell. Rhythms, or waves, are what make nature alive. Rhythms are the nature of nature.

Likewise, Irving Dardik came to his Superesonant Wavenergy Theory, which he now calls his SuperWave Principle, from nature, not from physics. It began, appropriately enough, with the heart. In March of 1985, Dardik's friend Jack Kelly, who was chairman of the U.S. Olympic Committee, took a morning run as he often did—and dropped dead from a heart attack. That unexpected event prompted Dardik to look closely at the activity of the heart.

Typically, the heart's activity is measured by the electrocardiogram, which shows repeated blips on a straight line graph, very linear. Dardik, a former vascular surgeon, studied it in a very different way. He saw that as an individual exercises briefly and recovers, his or her heart rate goes up and comes down: a simple wave of exertion and recovery. Superimposed on that large wave are smaller waves of contraction and relaxation of the heart muscle, systole and diastole—the beating of the heart. Further superimposed on the wave of each heartbeat are waves of biochemical oscillations associated with contraction and relaxation. Now there are three waves, nested and waving within one another. Go deeper, to individual molecules and atoms: these too are oscillating, or waving—yet another level of waves waving within waves.

If we look at this nested set in a macro context we see that it is also nested within waves: the daily wave of the circadian rhythm, the monthly wave of the lunar cycle, and the cycles of the seasons. Dardik's SuperWave Principle says that nature is *all* waves, right down to regularly pulsating DNA and all the way up to pulsating galaxies and the birth of the universe; waves, not as a description of behavior, but as the *fundamental stuff of nature*.

A critical aspect of Dardik's emerging theory involved the findings of a Dutch physicist, Christiaan Huygens, the inventor of the pendulum clock. One day in 1665, Huygens wasn't feeling well, so he stayed in his room. Unable to read or do any form of work, he stared aimlessly at two clocks he had recently built, hanging side by side on the wall in front of him. To his surprise, he noticed that the two pendulums were swinging in perfect synchrony, and they continued to do so for hours, for as long as they were wound. Curious about why this should be, he got up and changed the swing of one of the pendulums so the two clocks were out of synchrony. Within half an hour, they were again swinging in perfect unison. Were they influencing each other, he pondered, perhaps through vibrations mediated through the wall? To test that idea, he placed one of the clocks on the opposite side of the room. Before very long, the pendulums were no longer swinging in synchrony. Huygens's serendipitous observation led eventually to the development of a new sub-branch of mathematics: the theory of coupled oscillators.

To Dardik, this phenomenon, for which he coined the term *superesonance*, offered a way for thinking about how order might emerge in nature. He imagined a wall with many pendulum clocks, all in synchrony. This wall would itself be like a clock that, along with other such wall-clocks, could be mounted on a larger wall. These wall clocks would communicate with one another through sound waves and become synchronous, just as the individual clocks did. Dardik imagined this larger wall-clock being placed on a yet larger wall, together with other larger wall-clocks. As sound waves traversed the walls, overall synchrony would permeate the nested layers of clocks. Each clock would effectively be communicating with all the other clocks. (This thought experiment had its limits, not least in considering the practical issues of keeping the clocks maintained and running properly.)

Dardik's next step was to think of a nested, living clock: the microscopic cells of a living organism. In this scenario, the human body can be seen as a wall collectively composed of living clocks, the individual cells. Instead of resonating with waves of sound, they would resonate with the waves of biochemical oscillations. Dardik coined the term *wavenergy* for all such waves in nature. The wavenergies of all the individual cells would create a wavenergy of the organism as a whole, through superesonance, from the bottom up. This stronger wavenergy of the whole organism would, in its turn, help organize and synchronize the weaker wavenergies of the individual cells, again through superesonance, from the top down. The whole organism and the individual cells are engaged in a dynamic dialogue, mediated by wavenergy.

The wavenergy of an individual organism—a human, say—influences the behavior of other individuals. We often speak of a person's presence as his or her "energy," and we've all experienced the magic of "clicking" with someone in conversation, when our individual energies meld and something new and exciting emerges. That's an example of superesonance producing a new wavenergy we can *feel* in our own lives. So, too, is the emergence of "team spirit" at work. A small group of people come together to tackle a common problem, bringing their different skills and their different energies. When the team members click, a palpable new energy, or wavenergy, emerges, one property of which is the collective creativity of the team.

Something similar happens in ecological communities. We've all heard of "survival of the fittest," which involves competition among individuals. Biologists once assumed that if a species tries to invade a new community, and it is superior to a similar species within the community, it will successfully oust the incumbent species. As an ecological community matures and becomes established, it takes on properties *as a whole*. One such property is that in the above example, the would-be invader is repelled, even though it is competitively superior to its counterpart species. The cohesion of the community and its resistance to invasion is an emergent property, a new wavenergy through superesonance.

Similarly, local ecological communities, including humans, influence each other, producing a new wavenergy at a higher hierarchical level . . . and so on at ever-higher hierarchical levels, displaying ever more complexity through superesonance, so that the global ecological community, and the Earth itself, displays a global wavenergy. In Dardik's perspective, it doesn't stop there but continues onward and upward, to include the solar system, our galaxy, and eventually the universe as a whole. Top-down and bottom-up influences of wavenergy operate between and through all levels, a process that Dardik originally called *superlooping* but now terms SuperWaves.

Dardik views wavenergy as the organizing principle of the universe; through it, everything is connected to everything else. The whole and the parts are unified—what Dardik calls *matterspacetime*, including the living and the nonliving parts of the universe. This is in contrast with traditional science's view of matter being discontinuous and separate from space and time. Also part of the SuperWave perspective is the realization that the universe is not uncertain and chaotic but coherent and ordered, mediated by wavenergy through SuperWaves.

Thus, the theory has implications at the micro level, the macro level, and every level in between. Consider two important examples in the realm of physics. First, an example at the micro level, which relates to what quantum physicists call wave/particle duality. In 1803, the English physicist Thomas Young appeared to solve an age-old problem: What is the nature of light? Is it a wave, or is it a particle? With a very simple experiment, he "proved" that light is a wave. That satisfied everyone until a century later, when, in 1905, using data from a different kind of experiment, Einstein "proved" that light is in fact made of particles, which he called photons. That conundrum was settled in the 1920s by the French physicist Louis de Broglie, who said that light was *both* a wave and a particle, depending on how it is detected during an experiment. He went further and said that *everything* can exist as a wave or a particle. You don't notice the waviness of the chair you are sitting in because it has so much mass.

This extraordinary split personality of nature is shown clearly in the so-called double slit experiment, which appears to show that a particle, such as an electron or a photon, can be both particle and wave. The experiment, whose roots go back many centuries to Aristotle and Ptolemy, is a classic in physics. A light source fires photons at a solid screen that has two slits, A and B, side by side. Behind this screen is a second screen, which detects the pattern of light after it has passed through the slits. If slit B is closed when the photons are fired, the receiving screen records the light in the shape of the slit, just as if the light had arrived as photons, or particles, which is how they started out. The same thing happens when slit A is closed and B is open.

Weirdness happens when *both* slits are open. Instead of two overlapping images of the slits, which is what would be seen if the light was traveling as photons, the image is of alternating light and dark strips, like a pedestrian crossing. This is what happens when two waves *interfere* with each other, and it is known as an "interference pattern." Clearly, in this case, the light is traveling as waves after passing through the slits, when they started out as particles, or photons. The same thing happens when a beam of electrons is used instead of light: with one slit open, the electrons behave as particles; with both slits open, they behave as waves. This observation is problematic for classical physics because electrons must be either waves *or* particles and not change character en route.

This problem is exacerbated when a high-frequency laser beam is shone across the beams of electrons after they have passed through the slits. The result is that the pattern produced with two slits open is what would be expected if the electrons behaved like particles. Namely, there is no wavelike interference pattern but, rather, two overlapping images. Quantum physicists' explanation for this phenomenon is that because the light beam enables the observer to detect the position of each electron, this act of observation—the so-called collapse of the Schrödinger wave function—then determines the subsequent behavior of the electron as a particle. But exactly *why* this should be so, says Alan Lightman, "continues to baffle the best physicists in the world." According to the late, great Richard Feynman, "It is impossible, absolutely impossible to explain [it] in any classical way." The SuperWave Principle offers an explanation.

According to the SuperWave Principle, when the wave swarm, wave packet, or wavicle we describe as an electron is part of an atom, it is nested within the larger waves of the nucleus and thus maintains its coherence. When it is pulled from that context, however, it starts to lose coherence as its wavicle form disperses. What once looked and behaved like a particle now looks and behaves more like a wave. This perspective explains why an interference pattern is produced when both slits are open, because the electrons have all the characteristic of waves at this point. When illuminated by a high-frequency laser beam, however, the waveform is enhanced in frequency and amplitude, or coherence, and the electrons therefore behave like particles, as they do in atoms. At this micro level, what we recognize as a particle is in fact the peak of a tightly coherent wave.

The second example is at the macro level, where physicists face a problem when looking at the visible universe. When using the Newtonian equations of mass to understand galaxies and the universe as a whole, they come up with challenging conclusions. First, galaxies typically spin several times faster than would be implied by the observed total mass within them; so why don't they fly apart? Second, the universe is expanding far more slowly than would be predicted given the total mass in the universe—that is, there seems to be insufficient mass in the universe to manifest sufficient gravity to hold the universe together. By most estimates, based on Newton's laws of gravity and motion (the classical approach), there should be at least 10 times more mass (and as much as 100 times, by some estimates) in the universe for it to behave the way we observe it to do so. This has been called the "missing mass."

The proposed solution is that there is a vast amount of matter in galaxies and the universe as a whole that is currently beyond our means of detection. This missing mass has been given the name *dark matter*, because we cannot see it. Under the SuperWave Principle, the "it" does not exist at all. *There is no missing mass.* When physicists look at the mass in a galaxy and proclaim that it is insufficient to hold it together, they ignore the context of the galaxy—namely, galaxy clusters. The SuperWave Principle argues that the larger waves of the galaxy clusters maintain the coherence of the galaxies, so they don't fly apart. This is exactly equivalent to the larger wave of an atom maintaining the coherence

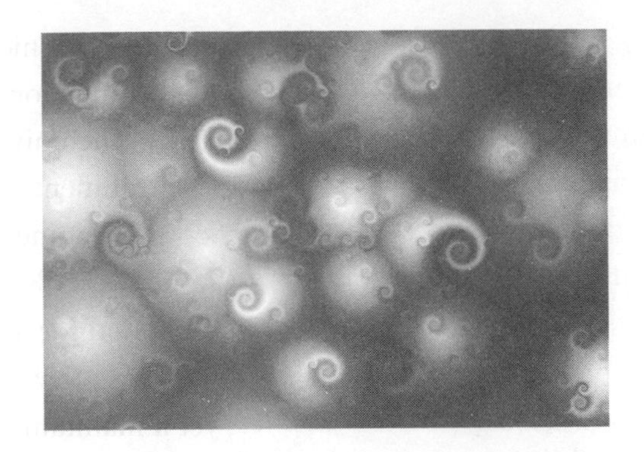

Top: oscillating chemicals in the Belousov-Zhabotinsky reaction, in a petri dish.

Middle: the spiral shape of a hurricane.

Bottom: the spiral shape of a galaxy. All three demonstrate waves waving within waves, or Superwaves: self-similar patterns (fractals) at different scales.

of the electrons, so they don't disperse as waves. This is a nice example of the fractal nature of Dardik's theory: the same phenomenon of coherence emerging from waves within waves, at all scales in the universe.

Is there empirical evidence to support this argument? Yes, and it is rooted in a 19th-century observation by a Scottish engineer, John Scott Russell. He noticed that when a barge moved up a canal, it generated a solitary wave ahead of it that traveled along the canal for long distances, without changing shape or speed. This solitary wave, these days called a soliton, is composed of many smaller waves, yet it maintains its coherence as a whole and over time. This apparently violates the second law of thermodynamics, which says that the collection of smaller waves that form the single wave should dissipate into its component parts. But the reality of the soliton demonstrates that the large wave has the ability to bring coherence to the collective of smaller waves, which in turn manifest as the large wave, a pattern of connection through relationship. Galaxy clusters and the universe as a whole can be seen as solitons.

The SuperWave Principle, Dardik argues, is a Theory of Everything. It is certainly a bold claim, and details of the theory can boggle the mind, but no more than some of the claims of quantum theory boggle the mind. A century ago the Danish physicist Niels Bohr said, "Anyone who is not shocked by quantum physics has not understood it." That sentiment remains valid today.

In the fall of 1989, Irv Dardik packed his car with boxes of books and papers, took leave of his family, and drove the 300 miles from his home in New Jersey to Pittsburgh to stay with his sister, Sylvia, who had agreed to help him get his theory down on paper.

Sylvia, the oldest of the Dardik children, is a sculptor and painter. She recognized the creativity that her brother has manifested since childhood, and that forged a strong bond between them. "Irv is an exciting person," says Sylvia, "and although my other brothers and sisters often thought he was off the wall, I loved his creative thinking. That's why he could move in with me for a couple of months. The others would have thrown him out!"

There was another reason Dardik chose to go to Pittsburgh, however, and that was a building: the Gothic-style Cathedral of Learning that rises majestically at the center of the University of Pittsburgh's campus. Located in the Oakland area of the city, not far from Sylvia's house, the building was erected in the 1930s and stands 535 feet high, making it the second-tallest education building in the world, and certainly one of the most exalting. "Walking into the main floor was like walking into a real cathedral," explains Dardik. "It is relaxing and inspiring, and I wanted a place like that to do my work." It was a grand setting for a grand theory to emerge.

Dardik went to the cathedral early each morning, when there were few students around, and spread his books out on several tables, just as he does at his home now. He made notes on a yellow pad in longhand and drew diagrams that wove together myriad ideas as the theory gradually, and graphically, took shape. In the evening, he went back to Sylvia's house, where the two siblings would have dinner. After that, they went upstairs to her study, and Irv would settle into a comfortable chair while Sylvia sat at the computer, typing as her brother read his notes. (Dardik wasn't computer literate then, and still isn't.) Often she would question what Dardik was saying and suggest better wording. "I was editing his text as I was entering it into the computer," explains Sylvia. "Irv has the most brilliant ideas, but he does have problems getting them down on paper. I'm an artist, not a scientist, but what he was saying made a lot of sense to me."

What Dardik was saying is simple and powerful. Science has long sought to understand what the universe is made of and how it works. In the 1630s, the philosopher René Descartes proposed that all things in the universe are automata, and the universe came to be viewed as a Great Machine. Three decades later, Isaac Newton's laws of motion and gravity explained how the universe works: as a "clockwork universe." Newton's laws allowed scientists to predict the movement of planets, which made possible the successful launching of spacecraft to rendezvous with our moon and, as happened in January 2005, a moon of Saturn, Titan. If everything in the universe followed Newton's laws, then everything would be predictable and essentially predetermined once it came into being and was set in motion. That's obviously not the case, particularly in the subatomic world.

In that strange land, it is impossible to know both the position and momentum of a particle, according to quantum mechanics, which entered the scene in the early decades of the 20th century. (This is known as Heisenberg's Uncertainty Principle.) It is therefore impossible to predict much about the behavior of particles. Indeed, a subatomic particle is not like a grain of sand, a thing, an object. Quantum mechanics views particles as "tendencies to exist" or "tendencies to happen." The quantum world is a world of probabilities and uncertainties, where particles may behave as waves, depending on the circumstances.

The quantum world and the world we experience day by day, where a table or a rock is something you can touch and doesn't change, could hardly be more different from each other. Yet, in Dardik's view, everything in the universe, from the smallest to the largest scale, are all manifestations of the same thing: waves. Dardik's theory therefore replaces the atomic theory of matter, which states that the universe is composed of particles and forces. No, he argues, *it's all waves*.

2

Humans and the HeartWave

"It is a wholesome and necessary thing for us to turn again to the earth and in the contemplation of her beauties to know of wonder and humility."

—RACHEL CARSON

FOR MOST OF OUR 100,000-YEAR HISTORY, humans thrived in small, nomadic bands, subsisting by hunting and gathering. The development of agriculture, the Neolithic Revolution, around 10,000 years ago, allowed people to establish permanent villages, with larger populations. But it wasn't until some 5,000 years ago that what we call civilization arose, in the form of city-states in Mesopotamia in the Middle East. City states such as Ur, Uruk, Legash, and Eridu were characterized by the emergence of specialized skills, such as pottery making and metallurgy, and the building of temples. These states were the harbingers of others to follow, such as Babylon and the Hittites, and then nation-states, like Greece and Rome. If you were to run a videotape through time, you would see a common pattern: A new center of civilization arises rapidly, dominating its geographical region; it maintains itself for a while, perhaps a century or two; and then it falls, rapidly, when another takes over.

The history of civilizations is a repeated wave of rise and fall, in every part of the world. In Mesoamerica and the Andes, you see the Olmecs, the Toltecs, the Aztecs, the Mayans, the Incas. They come and they go. Historians believe that there is something innate in the social and political dynamics of states that underlies this wavelike pattern. Often some kind of proximate cause is cited for the end of a particular civilization, such as drought a millennium ago that coincided with the disappearance of the Anasazi people in the American Southwest. But the persistent pattern of rise and fall is inescapable, whether drought occurs or not.

In more recent times, Portugal, Holland, Spain, France, and Great Britain enjoyed global dominance for a while, establishing themselves as global powers and forming colonies around the world, and then they retreated. It is as if civilizations march to the drumbeat of an inexorable dynamic of rise and fall, an inherent fundamental of human societies. The United States currently enjoys global dominance, ever since the Soviet Union collapsed. Hard as it is to imagine, this U.S. hegemony too will pass, as historian Paul Kennedy predicted in his 1989 book, *The Rise and Fall of Great Powers*, to be replaced by nations such as China or even India. Whichever nation assumes the mantle of global power, its time will pass. The inherent, wavelike dynamic of civilizations is inescapable.

Industrial corporations also rise and fall in wavelike manner. Arie de Geus, a former executive with Royal Dutch Shell, studied companies around the globe and found that most had a life expectancy of around 20 years, with larger, multinational companies lasting about twice that long. Very, very few lasted longer. In 1997, he published his results in the book *The Living Company*. The science of complexity has some insights into this birth-and-death phenomena of corporations (and of civilizations), which it views as complex adaptive systems. Such systems have their own internal dynamic, driven by—in the case of corporations—management style. And they interact with other complex systems, such as the market, suppliers, and competitors. Computer simulations of complex systems show very clearly that no matter how robust a system (a company or a civilization), its death is inevitable; myriad circumstances conspire against it in unpredictable ways.

The same wavelike dynamic plays out in the economic realm: the business cycle. When society is enjoying the fruits of a booming economy and low unemployment, it is often hard to see how or why the good times should end. But they do—always. Historically, Western economies have followed a pattern of boom followed by recession (or sometimes depression), an inescapable rhythm of high economic activity followed by low activity.

Economists have spilled acres of ink in treatises seeking escape from the business cycle, hoping to find ways of achieving constant high growth. But it is futile, because just as species and civilizations are complex adaptive systems, so too are economies. And with that recognition comes the acceptance that cycles of high activity interspersed with reduced activity are inherent to the system. Benjamin Franklin famously said, "The only two certainties in life are death and taxes." For economists, there's a third: the persistence of the business cycle. And just as extinctions in the biological realm set the stage for a burst of evolutionary creativity, recessions provide opportunities for bursts of innovation.

Life and death rhythms are inescapable in the nonhuman realm, too. For example, most species that have ever existed are now extinct; the average life span of a terrestrial species is around 2 million years. Like companies and civilizations, species are complex adaptive systems existing in a larger complex adaptive system, the ecosystem. And, again like companies and civilizations, species arise and then disappear, an inexorable rhythm of nature. *Homo sapiens* has been around for about 100,000 years, so we have a ways to go before we reach the average life span of terrestrial species. (Unless, of course, we so degrade the Earth that we bring about an early demise!)

Many people respond to the notion of natural rhythms positively and instinctively. It simply *feels* right. And where do rhythms feel more right than in music? Every culture has music, played or sung for a myriad of occasions, both happy or sad, or just for the fun of it. And even though music from other cultures may sound different to our ear—Eastern versus Western music, for instance—most forms of music are based on the octave. We respond to the rhythms of music by moving rhythmically ourselves, instinctively. Little children don't need to be told to sway, clap, or dance to music; they do it naturally.

Anthropologists now believe that the urge to produce and respond to music is wired in our brains. Why should that be? What survival benefit could it endow? Charles Darwin, who was always looking for evolutionary explanations of human capabilities, was puzzled by our musical bent. The musical faculty, he said, "must be ranked among the most mysterious with which [humans are] endowed." While anthropologists agree that playing and enjoying music is part of what it is to be human, they do not agree on the purpose for which it evolved. Some argue that it might have been important in courtship, while others suggest that group singing and chanting offered a kind of social glue.

Whatever the true explanation, there is no doubting that music making has been part of *Homo sapiens* throughout our history. Among the earliest known nonutilitarian objects, such as pendants and batons, found in the Paleolithic caves of Europe is a flute fashioned from the hollowed bone of an antelope some 43,000 years ago. It's not hard to imagine the music making, singing, and dancing that probably went on in front of the images of horses and bison in the prehistoric painted caves of Europe all those millennia ago. Someone once said, "Manners maketh man." But in truth, it is music that maketh man. Our resonance with rhythms is encoded in our genes.

Early in February 1985, Jack Kelly, brother of Hollywood legend Grace Kelly, who became Princess Grace of Monaco, was elected president of the U.S. Olympic Committee. A month later, in the early morning of March 2, a Saturday, Kelly jogged from his penthouse apartment in the Palace Hotel in Center City, Philadelphia, to the Fairmont Rowing Association's boathouse at 2 East River Drive. There he met seven friends who called themselves "The Golden Eight of Fairmont," a group of rowers with a cache of Olympic medals among them—hence their name. The men met every Tuesday, Thursday, and Saturday, just after daybreak, to propel their sleek sculls six miles along the Schuylkill River, a habit they had pursued for four years.

On this occasion, Mother Nature was smiling on the men, offering them a perfect winter day: a cloudless blue sky, warm air, and a river surface like glass; ideal conditions for sculling. Two boats were set on the water, both quads. Kelly joined Dietrich Rose, Sean Drea, and Bill Knecht in one boat, taking up the stroke seat at the stern, as was his habit. As stroke, Kelly's role was to set the pace, and that's the way he liked it. "Kelly was a tremendous competitor," says Al Wachlan, another member of the Golden Eight. "He didn't like to lose, even in practice." This Saturday morning's hour-long workout was no different, with Kelly setting a searing pace to get his boat out in front. At one point, one of his teammates pleaded for an easier pace. It was just a workout, after all. Kelly obliged, saying, "Is that slow enough for you?"

Back at the boathouse, the eight men stowed their boats and oars, took showers, collected their things, and prepared to leave. Kelly, now clad in a blue-and-white jogging suit, accompanied Rose to his car, chatting. Then he set out on his habitual postworkout run, waving to his friend and saying simply, "See you, D." That farewell would become especially poignant for Rose, who, 22 years earlier, had arrived in Philadelphia, knowing no one. Kelly had sponsored the young German's move from his native Berlin to the States and had waited at the train station to meet Rose and help him settle into his new life. With a firm handshake and a broad, friendly smile, Kelly was the first person to welcome the young immigrant to the city. Twenty-two years later, on that serene Saturday morning, Rose was to be the last person to speak to Kelly.

About an hour later, at 9:25, a passerby found Kelly, unconscious, lying on the sidewalk of the 1800 block of Callowhill Street. He was pronounced dead shortly after 10:00 A.M. at Hahnemann Hospital, but because he didn't carry any identification in his jogging suit, no one knew who he was. In a bizarre coincidence, Kelly's ex-brother-in-law, Eugene Conlan, collapsed and died on another street in Center City that same afternoon. Because the two men were still friends, Conlan carried a card saying that, in the event of an emergency, contact Jack Kelly, and gave the relevant information. The police searched in vain for hours for Kelly, not knowing that he was lying unidentified in the city morgue, where he had been taken from the hospital. It wasn't until 8:30 that evening that he was finally identified, fully 11 hours after he had stopped during his run, apparently to retie a loose shoelace, and died from heart failure.

That same night, Dardik learned of his friend's death from Bert Zarins, a member of the Olympic Sports Medicine Council and physician to the New England Patriots and the Boston Bruins. "I was crushed," remembers Dardik, "because we had been very close. But my immediate question was, why? Why would Jack, a man in prime physical condition, just die like that? It didn't make sense." A postmortem revealed that Kelly had some degree of arteriosclerosis, but so minimal as to be irrelevant as a cause of death. Dardik was determined to make sense of it, if only to understand what had happened to his friend. As it turned out, he came to understand much more than that.

Dardik remembered that eight months earlier, Jim Fixx had also died of heart failure after a run. Fixx's huge-selling 1977 book, *The Complete Book of Running*, helped launch America's fixation with fitness. Like Kelly, Fixx had some arteriosclerosis, but again so minimal as to be ruled out as the cause of death. Of course, the most famous endurance runner in history, Phidippides, suffered the same fate. Legend has it that in 490 BCE he ran the almost 26 miles from Marathon to Athens to declare the good news that the Persian military had been defeated. "Victory is ours!" he is reported to have said. "Victory is ours!" And then he collapsed and died of a heart attack.

Dardik delved into the medical literature and discovered that in the majority of cases of death associated with exercise, heart failure comes *after* the individual stops an activity, not *during* exertion, whether it is running, shoveling snow, skiing . . . anything.

The same year that Kelly died, exercise guru Kenneth Cooper, M.D., published *Running Without Fear: How to Reduce the Risk of Heart Attack and Sudden Death During Aerobic Exercise*. He argued that people who stopped suddenly after exertion risked pooling of blood in the legs, which can trigger heart failure. He called it "the great cool down danger." Dardik was skeptical. "I mean, people have been running for thousands of years, and they didn't die like that," he argues. "I began to think that it must be something in the *way* people run now that causes heart failure after exertion." When he looked further into the scientific literature, he found that people in technologically primitive societies—

hunter-gatherers—don't do sustained running: they run in bursts. Conversations with anthropologists confirmed this.

In the animal realm, too, sustained exertion is the exception. "Cheetahs, the fastest creatures on Earth, run in short bursts," notes Dardik. "And I found that all animals do that: a burst of exertion, then rest; a burst of exertion, then rest. You see this in whales when they dive, sharks when they hunt, birds in flight. Everywhere I looked, no matter what kind of animal was involved, it was the same pattern: a burst of exertion, followed by rest, and so on. You see it in young children, as well, when they are playing. They run. They stop. They run. They stop. This mode of exertion, which seems to be nature's way, these kids do it naturally—until we start training them in athletics, that is."

Dardik set out to discover what the heart is doing during exercise, particularly during intermittent exercise. "I strapped a heart rate monitor around my chest, put on a watch that displays the heart rate, and started running," he recalls. "I went from a resting heart rate of below 70 to up around 150. I kept running for 10 minutes, 20 minutes, and then stopped. My heart rate then started to recover, slowly. It took a while to get back to the resting rate." The picture was quite different with intermittent exertion: "It would go up to 150 rapidly, and then when I stopped, it came right back down to a little below the resting rate. I did this repeatedly—the same thing. I called it the pendulum effect. No one had seen that before."

Dardik began to name this pattern *waves of exertion and recovery.* When someone exercises, the heart rate climbs. When activity stops, heart rate comes back down. Repeatedly. What about the heart itself? When the heart contracts, that's exertion; and then it relaxes, or recovers. With every heartbeat, there's a wave of biochemical exertion and biochemical recovery; multiple biochemical waves on top of the larger wave of exercise and recovery, running and stopping. "I saw it as *waves waving within waves*," says Dardik. The phrase would become his mantra.

"I started thinking about the way we record heart activity in a doctor's office," says Dardik. "It's the electrocardiogram, intermittent spikes of activity on an otherwise straight line. But nature's not like that.

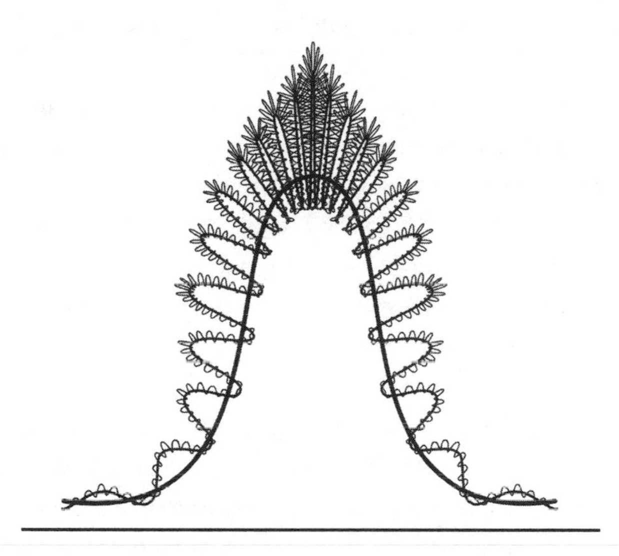

The HeartWave is a combination of the body's energy output during brief exercise and recovery, the large wave, and a representation of the heart contracting and relaxing (systole and diastole), the small wave superimposed on the larger wave. The third set of waves, the smallest, represents the biochemical oscillations of the heart function. This is not a conventional graph, with a single time dimension along the bottom. The bottom line is the time dimension for the large wave of exercise/recovery; the time dimension for the systole/diastole waves is the exercise-recovery wave; and the systole/diastole wave is the time dimension for the biochemical oscillations.

Think about when you exercise. Your heart rate rises. And when you rest, the heart rate falls. That's not a straight line. That's a wave! And when you plot the individual heartbeats along that wave, contracting, relaxing, contracting, relaxing, you have a series of little waves riding on the big wave of exercise and recovery. That's what I call the HeartWave." Dardik's key insight here is that someone who is doing cyclic exercise—exertion and rest, exertion and rest—is spending as much time training the body's recovery physiology as training his or her exertion physiology. "That's a new observation in sports medicine," says Dardik. "No one had thought about that. All they thought about was getting the heart rate up there and keeping it up there. And then, when you are finished, you keep moving, keep jogging, to cool off. God forbid that you should stop! God forbid that you should actually think about the physiological processes of recovery!

"I got a couple of triathletes with low resting heart rates, like Jack Kelly, put the monitor on them, and had them do a sprint and then sit down," explains Dardik. "Sometimes their heart rate would plummet from 150 down to 40, to 30, to 25. I said, 'My God, what is going on?' The heartbeat was in danger of flattening out, of stopping altogether. The pendulum effect had overshot in these people. Their blood pressure was plummeting, but these distance runners brought it on themselves because they were used to keeping on moving, of jogging when they finish sprinting. If you are a distance runner, you sure as hell have to keep running because you are at risk of dropping dead, because you have overtrained the stress side and not trained for recovery. It's the old 'use it or lose it' idea. The recovery side gets lost. So sometimes in athletes who haven't trained their recovery physiology, the heart rate hangs up there when they stop exercising, and sometimes they go into a steep pendulum effect and can't pull out of it. It's just like drug addicts who take heroin too much, too long, and then when they stop, they collapse. That's why people can be in danger of dying after sustained exertion. Their physiology simply isn't trained for recovery, so the heart rate doesn't make the upturn. That's what happened to Phidippides. That's what happened to Jim Fixx and all the others. And that's what happened to my friend Jack."

Some years earlier, when he was chairman of the U.S. Olympic Sports Medicine Committee, Dardik had felt in his gut that interval training rather than sustained exertion was natural and healthy. That intuition was now grounded in the physiological reality of the way the heart works, the insight about the importance of training recovery. And the image of the HeartWave—of waves waving within a larger wave—entranced him. It resonated with all the reading he had done in quantum mechanics, where waves are fundamental.

"Everything was coming together for me," Dardik recalls. "My work as a vascular surgeon, treating sick people. My work at the Olympics, where I had chronobiologists thinking about rhythms in the body and in athletic performance. My reading about waves in the quantum world, where waves are ubiquitous. I just had to figure out how it all fit together, what the larger pattern was. But I began to get the feeling that waves waving within waves had something to do with the unifying principle I was looking for. So I started looking for waves in nature and the universe."

3

Health and Harmony with Nature's Cycles

"The same stream of life that runs through my veins night and day runs through the world and dances in rhythmic measures."

—RABINDRANATH TAGORE

FOR THOSE OF US IN THE HIGHER LATITUDES, the seasonal rhythm of winter and summer is a part of our lives. People living in technologically simple societies followed the cycling of the seasons by marking the passage of the sun: the summer and winter solstices and the spring and fall equinoxes were times of celebration and preparation, dancing and singing. These same people followed seasonal rhythms naturally, too, not just in their hunting and gathering habits, but also in reproduction, and for good reason. Babies born in the summer, with warm temperatures and plentiful food to sustain the mother's lactation, had a better chance of survival.

In our modern society, of course, we have the means to dampen the climatic swings through air-conditioning and heating systems in our homes and vehicles. And food is in constant supply. Yet surveys show that in the northern hemisphere, most babies are born in the summer, with August being the most fecund month. This pattern of birth probably reflects cycles of sexual activity, spurred by seasonal cycles in male sex hormones in the fall, when days become shorter and sperm count is higher.

It's an inborn legacy of former times, unerasable by the trappings of modern living.

Moreover, babies born in the summer tend to be slightly heavier, which gives them an edge in the survival race. One curious fact, derived from a study of a large group of Austrian men, is that babies born in the spring tend to be slightly taller as adults. Why this should be is unclear, but it implies the existence of seasonal cycles in hormones that influence growth. Spring babies are conceived in the summer, when days are longest, but no one knows what the causal link might be, undeniable though the trend is.

Seasonal variation crops up in the likelihood of developing certain medical conditions, such as schizophrenia. The risk is highest for Northern Hemisphere babies born in February and March and lowest for August and September babies. Again, why this should be so is a mystery. Genetics is a factor in schizophrenia, but that cannot explain the seasonal pattern. It may be an environmental issue, such as greater likelihood of the pregnant mother's exposure to flu virus in the winter. Or perhaps it is the result of inherent seasonal cycles in factors that influence brain development. No matter what the explanation, this example, along with the examples of babies' birth weight and subsequent adult height, shows that no matter how much modern society insulates itself from yearly climate cycles, some effects of inherent natural rhythms persist in all of us

Animals, of course, experience seasonal cycles in a state of nature. The great blue herons that nest, feed, and breed in the rivers of New England fly to Florida in the fall and return the following spring; a million wildebeest in Tanzania's Serengeti Plains migrate north to richer pastures in Kenya's Maasai Mara reserve in May and then return in November; monarch butterflies flock in huge clouds from their habitats in the United States and Canada to volcanic highlands near Mexico City and coastal regions of southern California in the fall, journeys of up to 2,500 miles. Nature is replete with examples of animals of all kinds following the seasonal call to migrate, an irresistible instinct in tune with the yearly rhythms of nature.

Red deer stags on the Isle of Rhum in Scotland vie with each other to mate with as many hinds as they can during the fall rutting season, cued by a surge of hormones and shortening days. The females grow longer coats through the winter and survive on seaweed, ready to give birth in the spring. They are doing what all animals do in the wild: shaping reproductive activity around the rhythm of the seasons to give their offspring the best chance of surviving and thriving.

As a boy growing up in an orthodox Jewish family in Long Branch, New Jersey, the young Irving Dardik was fascinated by the sight and sound of waves of the ocean, just a block away from his house on Second Avenue. Whenever he could, he would go down to the beach, squat on the sand under the boardwalk, and watch the rhythmic patterns of the Atlantic waves as they rolled ashore. At night, in bed, he lay awake listening to their mesmerizing sound. Six decades on, Dardik is still fascinated with waves, but of a very different kind. He has a massive collection of magazine references to waves in all realms.

We see rhythms in the cyclic rise and fall through the 24-hour day of, for instance, blood pressure, body temperature, enzyme activity, hormone output, mental acuity, physical abilities. Menstrual cycles and testosterone levels cycle with the lunar month. Reproduction, migration, and hibernation cycles track the yearly seasons. Local ecosystems are a symphony of cycles in close resonance with one another, as is the global ecosystem, which James Lovelock and Lynn Margulis characterized as Gaia, the Earth Goddess. Even our own galaxy, the Milky Way, pulsates rhythmically, pushing out huge volumes of gas and then sucking it back in, a manifestation of its continued evolution.

Our own ancestors, the first members of *Homo sapiens*, arose in Africa some 100,000 years ago. They lived in harmony with the natural rhythms around them, hunting and resting, waking and sleeping in tune with nature. Before very long, populations of these people were moving into the Old World, Australasia, and, eventually, the New World, the Americas. No longer solely creatures of the tropics, human populations in a dizzying diversity of environments found themselves adapting to the novelty of seasonality and extremes of heat and cold that their ancestors had not faced. Humans, quintessentially, are adaptable creatures; but there are limits, of course.

Our hunter-gatherer forebears lived lives that we might judge as arduous: the food quest was unending; they were constantly on the move, living in small, nomadic bands; and they had few material comforts. However, anthropologists now know from studies of some of the rapidly disappearing groups of foraging people in today's world that they do not find their lot terribly difficult; they often enjoy leisure time that would be the envy of any Westerner. Anthropologist Richard Lee described the !Kung San of Botswana as working "bankers' hours." Just three hours of foraging a day would fulfill their food quest. More significantly, though, in these people, and in the bones of foraging people long dead, there is a virtual absence of any sign of chronic disease, and this is not a fortuitous outcome of shorter life spans. The same is true of other animals living in their natural environment: very little chronic disease. It makes intuitive sense that, living in harmony with nature and shaped by millennia upon millennia of evolutionary adaptation, creatures whose daily, monthly, and yearly lives are faithful to nature's rhythms will enjoy good health in the absence of accident or virulent infection.

Contrast that with modern society, where many of us enjoy material plenty and have access to the best that today's medicine has to offer. We have opportunities for exercise of all kinds and shelves upon shelves of vitamins and other nutritional supplements. There are multibillion-dollar businesses thriving in both those realms. And yet chronic diseases are rife: hypertension, arthritis, diabetes, multiple sclerosis, depression; these and other chronic diseases, including cancer, represent the large majority of the current health-care load. And the epidemic worsens with each passing decade. We are like caged animals that, unlike their counterparts in the wild, also suffer many chronic diseases. Why? Why should diseases that are effectively absent in our ancestors be so prevalent today? Indulgent lifestyles are part of it, with too many of us eating too much of the wrong kinds of foods. Stress is part of it, too, with 24/7 work lives driving so many people as they strive to amass yet more material trappings, or simply keep their jobs, because it has become the expected norm.

But the term "stress" is, for Dardik, just a different way of expressing something more fundamental; namely, that the life so many of us lead has torn us from the natural rhythms that are so much part of our being as *Homo sapiens*. We get up with the alarm clock, eat on the run, live under artificial lighting; air-conditioning turns summer into spring; central heating makes winter into summer. Quite simply, we have become disconnected from the kind of creature that evolution shaped us into being. With our natural rhythms and cycles disrupted or, more precisely, flattened, because much of what we do squashes them, chronic disease is the inexorable outcome. Unlike bacterial, viral, or parasitic infections, chronic diseases are a manifestation of a person's bodily chemistry and physiology being somehow out of balance; and in the case of autoimmune diseases, a person's immune defenses turn traitor and attack his or her own organs, a gruesome internal civil war. Chronic diseases are diseases of civilization.

Dardik is not arguing that, in order to reverse the slide into the ever deeper quagmire of diseases of civilization, we should eschew the benefits of civilization. Living the simple lifestyles of our hunter-gatherer ancestors is not an option for us today. The core rationale of Dardik's cyclic exercise protocol is to help people reconnect with natural rhythms within the context of modern lives. Its aim is to bring us as close as possible to being in harmony with nature while living and working in the ways that most of us live and work; that is, in offices, in factories, traveling on business. Living in harmony with the rhythms of nature is not a new idea, of course. The rhythm of yin and yang was, in fact, one of the foundations of traditional Chinese medicine, and the basis of health in premodern societies.

Modern scientific studies show that aerobic exercise is good medicine, helping stave off certain chronic diseases, particularly cardiovascular diseases. But the aerobic exercise industry has taken this too far, pushing us to do ever longer, more strenuous activities. No pain, no gain—really?

A more tempered perspective comes from a recent Harvard study of 40,000 men, which showed that brief periods of intense exercise interspersed over a period of, say, half an hour has better health benefits than running at less intensity, but consistently, for the same amount of time. The Harvard approach can be described as a form of interval training, which became fashionable in sports several decades ago. Instead of running at a sustained pace, people doing interval training sprint for relatively short distances, running much more slowly in between these faster bursts. To the developers of interval training, the procedure seemed somehow more natural. And it is, to a point.

Dardik's cyclic exercise protocol is akin to interval training, in that it involves four or five short bursts of intense exercise, each followed by recovery, just three or four times a week. A key difference is that the recovery period involves stopping completely, not exercising at a slower rate. The cyclic exercise protocol is as focused on training for recovery as on training for exertion, something that is absent in all other forms of exercise; it is more finely tuned to the daily and monthly rhythms of nature; and it is tailored to each individual's needs and goals. Above all, the protocol is natural, tuning the human body to the way other animals act in nature. It is the kind of exertion and recovery that, says Dardik, we are genetically and evolutionarily wired to do.

In short, the cyclic exercise protocol aims to close the gap that the pressure of living in civilized society has excavated between nature and us. It restores the waves that have been obliterated by our modern lifestyles. And, in restoring natural waves in the body, Dardik believes, it restores health.

Linda recently opened a small apparel and gift store, the Comfort Zone, in the small town of Hull, Massachusetts, 20 miles outside of Boston on the South Shore. On the face of it, she seems to have chosen a stress-free life, far from corporate America—which is somewhat true. In addition to her own store, however, she runs several businesses in the same office complex, including a health center for women, and she is cranking up a business-consulting operation. Linda is the kind of individual people have in mind when they say, "If you want something done, ask a busy person."

Linda has had many careers, all of them high-powered. As an executive with Coca-Cola, she introduced Diet Coke to the market and was showered with all the top sales awards for her region. She was recruited by Kepper Financial, in Chicago, where she spent seven years. During that time she was divorced, remarried, divorced; developed anorexia; and was a single mother. AIG, in New York, then recruited her as head of a division that establishes off-shore mutual funds. Fleet Bank, in Boston, then stepped into the picture, where she was in charge of retail investments, to the tune of several billions of dollars. Tiring of corporate life, Linda left Fleet in 1995 and teamed up in a smaller business venture in New Jersey with Neale Godfrey, Irv Dardik's sister-in-law. It was a move that, Linda now says, saved her life.

Linda had had what she describes as "a strange childhood." As a small child, she lived with her father, a severe diabetic who went blind at the age of 27. He was dead at age 34, when Linda was just 12. She then spent several years in various foster homes, until her mother returned from England. An academically precocious child, she graduated from high school early. But, by her own description, she was an undisciplined teenager, into all the kinds of self-abuses that make parents' hair stand on end. For adventure, she took off for Richmond, Virginia, never once making contact with folks back home for a dozen years. At age 19 she started to get her life back, went to City College in Richmond, and studied business administration while working in a bank. It would be seven years before she graduated. That's when it was time for the beginning of corporate life. From the very start she had severe problems, quite apart from family issues and eating disorders and a bad self-image, as she describes:

"I had kidney problems, all kinds of problems for years. Finally, after two years of tests they said, 'Oh, you are diabetic.' I was 27 at the time, and it terrified me, because I had seen my father go blind and die from diabetes, so I knew how bad it could be. The condition continued to get worse. The doctors' solution was to give me more and more medication, but it didn't help. I have to admit that I wasn't a very good patient. I didn't always follow instructions, so I can't say it was all the fault of the medicine.

"By the time I was in New York I was on five medications, including antidepressants. I was suicidal at times. They gave me insulin on top of everything else, but it still didn't help. A1C tests, which tell you if your organs are being damaged by the disease, were high: 11 compared with 5, which is the normal number. I had laser surgery on one eye, for burst blood vessels. I had lost all feeling in my feet and suffered excruciating pain in my legs.

"One result of diabetes is that you find it hard to concentrate; you get tired. I remember sitting in meetings, digging my nails into the palms of my hands so I could remain focused. I had a hard time doing anything. I would get very low sometimes, and I would never know whether it was going to happen in meetings with clients. But I am a very determined individual, and I was very successful.

"When I met Irv, I was having all these physical issues, but I didn't talk to him about them for several years. We talked theories, concepts, but not my personal disease. But when I was told I was going to need a kidney transplant in a year or two, and all they were doing was trying to give me more medications, I thought, 'Why not give it a try with Irv?' He said, 'I was waiting for you to ask!' Sure, I was skeptical, but what was there to lose at this point?

"I started in 2001, going to Irv's place three or four times a week, doing the cycles program. Within a couple of weeks the pain in my legs started to lessen. I used to fall a lot because of the numbness in my feet, but now I wasn't falling as much. I used to have horrible nightmares, and they started to disappear. I would check my blood four times a day, and within a month the numbers had stabilized, and soon the sugar levels began to drop, although they were still above normal. After three months I decided I would stop taking my medications, but I didn't tell anyone because I didn't want the arguments. I also wanted to wait until I had had my numbers from my next checkup, which was six months after I started the cycles. It was amazing. The doctor in New Jersey said, 'Your blood sugar is fine; your A1C test is 5/6. You don't have diabetes!'

"I wanted to hear those numbers because I knew I was improving physically and mentally. I wasn't tired anymore. I was like a different person. I had energy. And my eyes were improving, something the doctors told me was impossible. They said, 'You will have to have laser surgery constantly.' That's not for me. I don't wear prescription glasses anymore. The scariest thing in my life right now is to think that I could go back to all that. I get mad at myself sometimes, when I don't stick to the exercise regimen properly. But I am determined not to let that happen.

When something like this has saved your life, you don't want to mess with it. The difficult thing about telling people about the program and its effects, in a funny way, is that it is so powerful. People just don't want to believe that so little exercise can have the impact that it does."

Dr. Bernard Hering, a surgeon at the University of Minnesota and a specialist in transplanting islets of Langerhans (insulin-producing cells) into diabetic patients, talked to Dardik about Linda's experience. "The course of recovery that she followed exactly tracked what you see in patients I give islet transplants to," he later told me. "I admire Dr. Dardik's work, but in the absence of a proper clinical trial, it is hard to say anything definitive about a single case, such as Linda's." Asked if the kind of reversal of symptoms Linda experienced is seen to happen spontaneously, however, he said, "No."

4

Bucking the System

"Is the system going to flatten you out and deny you your humanity, or are you going to be able to make use of the system to the attainment of human purposes?"

—JOSEPH CAMPBELL

JUST AS THE SUN'S SEASONAL ODYSSEY across the sky was well understood by technologically primitive people, so too was the moon's monthly (29.5 days) rhythm of waxing and waning absorbed into society's collective psyche. The full moon is traditionally a time of high energy and excitement, while the new moon is a time for quiet and reflection, all of which was taken very seriously. The Moche people, who lived in northern Peru some 4,000 years ago, conveyed equal recognition to both moon and sun: their cities were dominated by two imposing temples, one dedicated to the sun, the other to the moon, both quite elaborate. The sun was viewed as representing male energy, of domination, while the moon embodied the feminine, a nurturing energy. This was common lore was shared throughout Mesoamerica and South America, as well as in the Old World, although its expression took different forms. The waxing and waning of the moon was considered to be symbolic of the growth and decline of plant, animal, and human life, of rhythms of birth and death.

The moon's daily (actually, 24 hours, 49 minutes) orbiting of the Earth drives the rhythms of the tides, of course, engendering two high tides and two low tides each day through its gravitational pull. And twice a month, at full moon and new moon, the high tide is higher than usual (so-called spring tides), creating a 15-day cycle. The feeding and reproductive rhythms of countless marine and other coastal creatures slavishly follow the ceaseless ebb and flow of the tides, particularly the 15-day spring-tide cycle.

Even trees respond to the motion of the moon. Scientists in Italy and Switzerland discovered that the diameter of a tree stem fluctuates in synchrony with the tides. They suggest that the moon's position creates a miniature tidal flow that moves water between living cells and the trunk's structural framework of deadwood. This intriguing observation might explain widespread folklore urging that, to hasten drying of the wood, trees be cut before a new moon.

In the realm of human physiology, women's menstruation is the most striking example of a monthly rhythm. (Oddly, a 1996 survey by the American Medical Association showed that only three out of four doctors—mostly men—agreed that the menstrual cycle was a biological rhythm at all!) The preparation of the uterus for the implantation of a fertilized egg and the sloughing off of the uterine lining if implantation doesn't occur are orchestrated by interlocking waves of release of the hormones estrogen and progesterone. No one knows why the length of the menstrual cycle matches almost exactly that of the lunar cycle, but it would be an extraordinary coincidence, if coincidence it is. More likely is a tight coupling of women's reproductive physiology with the lunar cycle for fundamental, and mysterious, reasons.

In people living in harmony with nature, women usually ovulate at the full moon and experience menstrual bleeding at the new moon; men's production of testosterone peaks at the full moon, too, in phase with the time of ovulation. Women living in close proximity with each other, whether in hunter-gatherer bands or in dormitories, for example, come to ovulate in synchrony, creating a community rhythm in phase with the lunar rhythm.

Any other kid in second grade would have just done what the teacher asked. But not seven-year-old Irv Dardik, who even at that early age liked to do things differently. Simply producing a short account of his summer vacation—the first writing assignment of the school year—wasn't enough for him. He wanted to make it more interesting, so in what must have been a painstaking exercise, he used an array of colored pencils, each letter of every word a different color. "It took a long time," he now remembers, "but I thought it was extremely creative, pretty. I thought the teacher would like it."

Not only did the teacher not like it, he was outraged. He thought the boy was trying to be a wise guy. "I was taken by the arm, dragged to the principal's office on the second floor," Dardik recalls. "They plopped me down in a seat and called my parents. They said I was insubordinate, not following directions, that I was doing things to disturb the class." Dardik's father agreed with the teacher and severely reprimanded his son. "I was completely bewildered, and couldn't make sense of why I was being punished for doing something that took special effort and that I thought was terrific." That was the first of many times that the maverick Dardik would be dragged onto the carpet for thinking outside the box.

The incident apparently went deep into Dardik's psyche. A couple of years ago his wife, Alison, was having hypnotherapy sessions and, just to humor her, he went for one, too. He had a problem he thought the therapist might help him with: writer's block. Although his mind is a geyser of ideas, getting them down on paper is sheer agony, often to the point of impossibility. The therapist asked Dardik to close his eyes and began the hypnotic procedure, music playing softly in the background. "Then the guy tells me I can't open my eyes," Dardik says. "'Sure,' I thought, but I played along with him. All of a sudden I started talking. I said, 'I like my colors,' and the colored writing story came flooding out. I was shocked because I hadn't thought about it for six decades. It wasn't just retelling the story to this stranger. I *felt* it viscerally. I felt how good it was doing it, and how awful it was to experience the crushing reaction from the principal and from my father, who was a very tough guy. He gave me hell, and the principal threatened to expel me. It was like doing a painting you're really proud of and being told it is terrible."

Maybe the incident is the source of his writer's block; maybe not. In any case, even today Dardik likes his colors. He finds it impossible to read an article or a book without penning copious annotations in the margins and doing multiple underlining in a wild spectrum of colors. Dardik is an avaricious reader, mostly science of one kind or another, and his house is littered with piles of articles and books that look as if someone spilled a painter's palette over them.

Dardik's father, Morris, was born in the town of Berzin, near Minsk, White Russia, in 1901. When he was 13, his father died, and his last words to his young son urged him to flee the approaching guns of World War I, which he did. He ended up in Samara, a town on the Volga River in south central Russia, and in his late teens he joined the Russian army as a special messenger, traveling the country with a briefcase handcuffed to his wrist. Later he was a quartermaster, which gave him access to scarce food supplies, a temptation that got him into trouble and nearly cost him his life.

One day he went to the post office with a parcel for his mother. Suspicious of its contents, post office officials opened the parcel, revealing a cake. For this simple theft of army property, the young quartermaster was packed off to jail, a place in which few people survived. While he was being marched to what would have been a grisly fate, he saw a friend, the town's mayor, whom he persuaded to arrange for his release. He and his wife arranged to have themselves smuggled into Poland, where they boarded a ship in what is now Gdansk and sailed for the United States.

They settled first in a modest house in the small town of Spotswood, New Jersey, but soon moved to an apartment in Long Branch, a town that once was one of the earliest and most glamorous seaside resorts in the country. In the latter part of the 19th century and for a few years into the 20th, it was a favored summer spot for the rich and famous, including seven presidents. By the 1920s, however, the town's glorious reputation and buildings were in decline. Later, in the 1960s, organized crime made

Long Branch their base, and a spate of gangland murders wiped out whatever good standing the town retained.

By the time the Dardiks arrived there in 1922, Mr. Dardik was suffering from a severe form of arthritis, ankylosing spondylitis. He opted for an experimental form of surgery at New York's Hospital for Joint Diseases and was in a body cast for months. The outcome was not at all what had been promised, and he finished up with his spine and his hips completely fused. As a result, he walked painfully and slowly, profoundly stooped, steadying himself with a cane. Mr. Dardik found ways around his crippled physical condition, however, such as using a stick with a nail attached at the end to pull up his socks (though he needed help with his shoes). He even managed to paint the walls and ceilings in the family's home, using his arms to pull himself up a stepladder while his oldest son held it steady.

Despite his physical limitations—or maybe because of them—Mr. Dardik was an aggressive, domineering man, very authoritarian, and he maintained a strict orthodox Jewish household. Rules of Sabbath were strictly followed, and Hebrew lessons and Bible study were part of everyday life for his three sons and, to a lesser extent, his three daughters, who were all older than the boys. The language of the home was Yiddish, and Irving didn't learn English until he went to school at age five.

Dardik's mother, Sarah, was a sweet, gentle soul who never had a cross word for anyone. She was, however, very much under the thumb of her husband, who loved her dearly but never let her out of his sight socially, as a way of controlling her. When she died, the eldest son, Elliot, picked out a phrase from one of the psalms as an epitaph on her gravestone. Translated from Hebrew, it reads: "Her ways were ways of pleasantness, and all her actions were for peace."

The father helped support the family by giving Hebrew lessons and by slaughtering chickens in the basement, following kosher rules that generations in his family had hewn to for centuries. His wife worked long hours as a seamstress in a nearby sweatshop. As is often the case with poor, immigrant Jewish families, there was an expectation that the children—particularly the sons—would do well academically and become doctors. Everything the parents did, financially and through guidance and cajoling, was for the sons to achieve this end. They did.

By the time Irving was born in 1936, the Dardiks had moved to a Victorian house, 62 Second Avenue, just down the street from their apartment and one block away from the ocean. It was a modest enough house: two stories, a basement, small yard back and front, and a porch front and side. With four bedrooms and six children, there was a lot of bedroom sharing. The family was close-knit, eating together, going to synagogue together, playing card games and chess on Saturdays. But having fun was not part of the equation. Irv Dardik, the youngest of the children, found it all extremely stifling. "Every day was the same thing," he recalls. "You get up, you have to study the Bible in the morning. The same packed school lunch, a boiled egg and the same sandwich. The same dinner for each day of the week, soup and chicken every Friday. My parents were terrible cooks! Going to synagogue, the same prayers. I couldn't stand the rituals, the monotony."

Dardik detested rote learning, no matter what it was: Hebrew, school lessons—anything that required following rules. He felt caged in, physically and mentally, in the classroom; he was endlessly curious and much preferred to figure out things for himself rather than being forced to ingest what he was given. This same trait continued through university, medical school, and beyond.

Dardik found many ways of escape. Some were as simple as mumbling in synagogue, instead of reciting the words of the designated prayers. Often he would skip classes and go to the ocean, where he ran along the beach or hid under the boardwalk, watching and listening to the waves. Waves entranced him. Motion fascinated him.

Hiding he did a lot, and he had favorite spots: in the branches of the cherry tree in his backyard, under the porch, and in the attic, where he devoured his beloved comics such as *Captain Marvel*, *Blondie*, and *Superman*. There was a lake not far from his home, another favorite sneaking-off destination, where he loved to be out in the open. From an early age, Dardik was very much a nature boy, luxuriating in the sights, the sounds, the feel of the wind on his face. It made him feel whole. Even though sneaking off was strictly against the rules, at school and at home, he seldom got in trouble over it. As the youngest in the family, Dardik got favored treatment, allowing him to get away with things his siblings wouldn't even dare dream of doing. Not surprisingly, they hated him for that, and that feeling simmered for many years. On the other hand, Dardik's parents came to love and respect their youngest son's sense of rebellion.

Dardik also discovered he had a natural talent: running. He ran a lot, for the pure joy of it, on the beach and by the lake, but he knew he was good. One summer while he was on the beach, Dardik saw his school's track coach, who was also a lifeguard, working some of his runners. Confident in his running abilities, Dardik boasted, "I can beat any of your guys on the track team." The coach just laughed, and Dardik felt brushed off. "I felt slighted," he now says, "because, I don't know why, but I sensed I had an unusual talent, and the coach was ignoring me. And that, I thought, was that." The following school year, 10th grade for Dardik, he found himself assigned to the coach's homeroom. One day early in the semester, the coach said out of the blue, "Do you still think you can beat my guys?" "Sure I do," said Dardik, surprised but delighted. "OK, we'll give you a chance," the coach replied.

The next day the coach paired Dardik with Howie Limner, the state-ranked half-miler, and had them run a lap and a half, 660 yards. "I stayed shoulder to shoulder with him the whole time," says Dardik, "and then, coming off the last turn, I passed him. At the finish I fell flat on my face, and I spent days getting cinders out of my skin."

No one in the family had ever dreamed of being involved with sports, so Dardik was a little nervous when he told his father what had happened. "When he heard that I had won, he was my biggest supporter ever after," remembers Dardik. Subsequently, as captain of the school track team, Dardik the sprinter attended meets on weekends, which often took him out of his hometown and exposed him to people outside of the community that was so central to his family's daily life. "All of a sudden, I felt recognized," says Dardik, "instead of being hidden away in a Jewish community with a couple of Jewish friends at school." He was soon capturing state and national titles, and he eventually established himself as an Olympic-class sprinter, the source of an agonizing decision he would have to make in his early 20s: would he be an Olympic sprinter, or would he go to medical school?

Other talents came easily to Dardik, too. He was a kid who could turn his hand to anything and excel. Elliot, Dardik's eldest brother, describes his brother this way: "He was very gifted. When he was three, four years old we used to play card games on Saturdays. One had a hundred different kinds of birds. He knew every bird. Another game had pictures of classic paintings—Renoirs, Monets, that kind of thing. He knew them all. We also played chess together, and he often beat me, even though I am six years older. He taught himself to play the trumpet and learned Haydn's trumpet concerto, just by listening to a record. He started piano lessons, like we all did, but soon gave them up—couldn't stand it. But then he taught himself and played brilliantly. He can't read music, but he composes beautiful pieces, just sitting there at the piano. He's quite a guy, and he's always been there for everybody. One time my other brother, Herbert, was being bullied by this big guy at school. Irv didn't think twice about it; he just went out and beat the shit out of this guy, even though he was bigger than Irv."

Even though he hated learning the conventional way, Dardik loved reading, not just comics but Bible stories, too. One especially had a powerful impact on the young Dardik. It was a story about King David of Israel. "I was just three or four years old, still sleeping in a bed in my parent's room," Dardik recalls. "I had a comic book of Bible stories, the kind for kids, thick, printed on cardboard. The story I read over and over was when David was running for his life, with King Saul and his soldiers searching everywhere for him, bent on killing him. David hid in a cave, and the soldiers were closing in on him. God sent a spider to the cave, and it quickly wove a thick web over the entrance. When the soldiers came to his cave and saw it was covered with a spider's web, they moved straight past, not imagining that the web was freshly made. And David was safe.

"The story fascinated me. I used to lay awake at night, wondering what it was like to be David, hiding in the cave, safe. Then one night I woke up suddenly, with this strange feeling that I was somehow connected to David and that, like him, I had something important to do in my life. I had no idea what it was, but I knew there was something I had to do." Hiding, like David, or "caving," as he now calls it, is core to what

Dardik has been for much of his life: a man who seeks a safe haven so that he can do whatever it is he wants to do, outside the system.

The feeling the he had something important to do in the world remained with Dardik through high school, through university and medical school, and through his career as a vascular surgeon. But he held it in a sacred, secret place in his heart. He told no one, not his parents or his siblings, not his friends or even his first wife, Sheila. He told no one until he met Alison Godfrey, in 1979.

Despite his repeated skipping of class at high school and learning in his own, unorthodox manner, Dardik did well enough to go to college, but not well enough to earn an academic scholarship. He faced the prospect of having to work to pay for tuition, just as his brothers and sisters did. But that wasn't his greatest concern. His greatest concern was what to do there. Nature was his passion. Running was his passion. But ever since he was a kid, there was the expectation that he would go to medical school and become a doctor, a healer, a financially successful individual in the Jewish community. Elliot, the eldest son, was already at the University of Pennsylvania, on the way to becoming an oral surgeon. Herbert was also on track to becoming a vascular surgeon, at New York University Medical School.

More out of a sense of pleasing his father than following his heart, Dardik applied to Penn, to do a premed degree. He was admitted and entered in the fall of 1954, rooming with Elliot for two years, first in an apartment on 40th Street and Pine, around the corner from the dental school, and then on Chestnut Street, in a house owned by an Armenian family. "We had to leave the first apartment because the bedbugs were so awful," remembers Elliot.

A few weeks into the first term at Penn, Dardik's high school principal called Dean Robert Pitt and told him of Dardik's prowess in track, and the dean promptly awarded Dardik a full academic scholarship.

Dardik hadn't mentioned his running achievements on his application. He was now free of money worries, but not of the anguish over what he should do with his life. He became cocaptain of the track team, running times that put him in national ranking. He ran internationally and won a gold medal in the 400-meter sprint at the 1957 Maccabiah games in Israel, a kind of Jewish Olympics, which, he says, "made me the fastest Jew in the world!"

All of this he loved. But there was a dark side. At the beginning of his second year, Dardik sank into an existential crisis. Barely scraping by academically, he was asking himself, "Why am I doing this? I'm going down the wrong track. What's my passion?" Dardik went to Pitt and said, "I can't do this." They talked for a long time. Pitt was sympathetic to Dardik's soul-searching and suggested he take a semester off to think things through. He did, though it wasn't particularly satisfactory.

"I just hung out a lot," recalls Dardik. "My parents didn't know I was taking a semester off. I spent a lot of time thinking about where I was going, but I couldn't think clearly. I couldn't see a different path to follow, not with any deep conviction, anyway. I was pretty depressed, and there was no track to divert me because it was winter. I didn't want to be a doctor, but what else does a young Jewish boy do? I knew I was doing it for my parents, particularly for my father. When the next semester came along I felt I had to go back, but not with much enthusiasm. At least it was track season and I could start running again." Shortly afterward, Dardik met Sheila, whom he later married. Being in a committed relationship compelled Dardik to knuckle down and finally do what it appeared he was meant to do.

During senior year, Dardik took time off again, this time involuntarily. He came down with mononucleosis, a not-so-subtle sign that he was putting himself in a place where his psyche knew he shouldn't be. A practical consequence of that lost semester was that when Dardik finished at Penn in 1958, he wasn't qualified to graduate because he hadn't fulfilled the course's time requirements. Even so, he managed to persuade the faculty at Hahnemann Medical College in Philadelphia to accept him. The decision to go down that path had again been anguishing, this time for practical reasons. "I knew I had a good shot at making the Olympic team," Dardik explains, "but I was told that if I did that, between premed and medical school, I would never be able to get back into any medical school. They said, 'If you take off for track, it means you aren't committed.' It was a tough decision because I had just been competing in a European tour. My last race was my best, 20.9 seconds for the 200 meters. But I conformed and opted for Hahnemann, kissing good-bye any chance of walking into the Olympic stadium as an athlete."

Thus began what he now describes as "the four least enjoyable years in my life." Dardik felt himself encaged again, subjected to rote learning, frequent tests, a toxic competitive atmosphere, and a course that to him was at odds with learning how to be a healer. "You go to med school to learn about life, and you start with a dead body to dissect," he says with obvious pain. "That to me is bizarre. This is supposed to teach me what health and disease is all about? I wanted to learn about the wholeness of health, and all the course did was chop it into little bits."

There were 200 students in anatomy class, held in a huge dissecting room, with four students to a cadaver. "It was very competitive," says Dardik. "Everyone knew that only half of them would graduate, so there was no sense of people helping each other out. I was trying to understand anatomy, what it looked like in life. If I went to look at someone else's dissection, to compare it with mine, they would cover it up to stop me from seeing what they were doing." To get around this, Dardik started to go to the dissecting room at night when no one else was around. "I befriended the guards so they would let me in. I would look at all the dissections and look for patterns and variations. I learned a lot that way," he says.

Just as he had at high school, Dardik started skipping classes—only more so. He went to none. Instead, he went around to the bookstores of the other medical schools in Philadelphia and bought the required textbooks, which were different at each school. "I would take them back to my room, lay them all out on the floor, and see what they were saying about the same things," says Dardik. "So I was going to five different medical schools at once!" He knew he was learning what he needed to know, but doing it in his own way. As the anatomy year was coming to an end, Dardik was summoned to Dean Cameron's office. "I hear you don't attend classes, or go to lectures or to the lab, and you don't write papers," he told Dardik. "Can you give me a good reason why you should be allowed to take the final exams?" Dardik explained his nocturnal comparative anatomy habits, his self-tutoring with all the different medical texts, and the dean said, "OK, I'm curious. Take the exams." He passed with flying colors, much to the chagrin of his classmates, who despised him. They didn't like the fact that he hadn't graduated from Penn, yet was allowed on the course. They didn't like that he skipped classes and lectures, didn't socialize with them, and spent an inordinate amount of time at track. Now here he was sailing through the exams. Once again, Dardik was getting away with bucking the system.

Part of that way of learning included talking with the professors. "I loved to do that, asking questions, not just accepting the dogma in the books," says Dardik. "It's the way I learn best, through talking." It was also the way that he finally got his degree from Penn. He spent a lot of time talking with the physiology professor, who took an interest in Dardik and his ideas. Pretty soon Dardik was helping the professor with a research project. The professor was impressed enough with Dardik's skills and insights that he called Dean Pitt at Penn and told him about it. Pitt promptly called Dardik and said, "You can have your degree." That was in 1958, two years after he had embarked on his medical degree.

Elliot remembers a time when Dardik's habit of not going to class nearly caught up with him: "One Friday in his first year, Irv remembered he had a test on Monday morning, cat anatomy. The professor would stick pins in a dissected cat, and you had to identify the different anatomical structures. Irv hadn't been to class, so he didn't know cat anatomy from beans. He called our brother, Herbert, in New York, and asked him to come down to help him. The two of them spent Saturday and Sunday dissecting a cat, going over the anatomy, learning it. And then Irv goes and gets an A on the test on Monday. I don't know how he got away with this stuff!"

By the end of the four-year MD course at Hahnemann, Dardik was miserable. He was disillusioned with medicine. He had stopped running. And he started to get sick, with pains down the outside of his legs. When he graduated, Dardik signed up for his internship at Hahnemann, and very soon he got even sicker. He had a high fever, the pains in his legs were worse; he had pains in his neck and in his ribs that were like knife jabs every time he inhaled. He also had diarrhea and bloody lesions on his legs.

"I was hospitalized for six weeks, felt horrible, could barely get out of bed, but no one knew what it was," says Dardik. "I had recently given mouth-to-mouth resuscitation to a woman who had tuberculosis and subsequently died, so they thought I might have contracted TB from her. They wanted to give me cortisone because of the inflammation, but if you have an infection such as TB, which they suspected, and you take steroids, you are dead. Eventually, when they couldn't find any indication of TB, they gave me intravenous cortisone. Instantly I was much better. I was back in the world. When I got out of hospital I went to the track just over from the hospital, climbed the fence, and ran—well, crawled really, around the track, on my hands and knees, I was so weak from the illness. I said to myself, 'I am never going to stop running again.'" It wasn't until a year later that he got a diagnosis: ankylosing spondylitis, the disease that had crippled his father.

Dardik was accepted for a post in internal medicine residency at Hahnemann. He had no interest in surgery and was determined to work with patients as a physician, talk with them and be a source of comfort to them, unlike the aloof doctors he had seen at Hahnemann. Just before he was due to take up that position, Dardik visited his brother Herbert, who was a surgical resident at Montefiore Hospital in the Bronx, New York. "After doing rounds, Herbert introduced me to the chief of surgery, Elliot Hurwitt, a man of international renown. We talked for a few minutes. He said, 'Come to my office.' He sat down and said, 'I am offering you a position as a surgical resident here.' I said, 'You know I have a residency at Hahnemann.' He said, 'That's OK.' I accepted the offer on the spot. I guess he greatly respected Herbert's work," says Dardik, "and he thought having two Dardiks as surgical residents would be good for the hospital.

"Even at the time it was hard for me to explain in rational terms why my answer was yes," says Dardik. "It was one of those gut feelings you sometimes have and you trust it completely; that's how it seemed at the time. Yes, I had had enough of Philadelphia. And, yes, I had grown tired of the aloof and regimented atmosphere at Hahnemann and with the way

I was treated, with disdain really. But it had never occurred to me to say to myself, 'OK, you've got to get out of here; go be a physician at some other hospital,' let alone go into surgery. And yet when Hurwitt invited me to Montefiore to be a surgical resident, I immediately knew that accepting the offer was my only option. It was as if the whole thing was beyond my control and I just had to trust it: 'OK, so you are going to be a surgeon, even though you never wanted to be one.' I also knew it was a risky decision, because I was imperiling my career by going back on my promise to be a resident at Hahnemann. When I went back to Hahnemann and told them I was backing out of my commitment, they were definitely very unhappy about it.

"At Montefiore, I immediately felt free; I could breathe again for the first time in years. I felt I had been released from a cage."

5

Bucking the System, Again

"To accomplish great things, we must not only act, but also dream; not only plan, but also believe."

—ANATOLE FRANCE

LET'S TALK ABOUT PLANET EARTH, as a whole entity. Life thrives everywhere, even in some of the most unlikely and inhospitable places, such as in Arctic frigidity and in infernos in underwater volcanoes. The diversity of life-forms on Earth is remarkable, even wondrous, what Charles Darwin described as having "grandeur." Even more remarkable is the harmonious interplay among the countless populations of organisms and the physical environment they inhabit.

In the 1970s, chemist James Lovelock and biologist Lynn Margulis provoked near outrage among scientists when they proposed that the world is not simply populated by organisms that interact with one another. They suggested that the planet is itself a living organism, a proposal known as the Gaia hypothesis, named for the Greek goddess, Mother Earth. Lovelock defines Gaia "as a complex entity involving the Earth's biosphere, atmosphere, oceans, and soil; the totality constituting a feedback or cybernetic system which seeks an optimal physical and chemical environment for life on this planet."

To many scientists, the idea that our planet was tending to itself, nurturing itself in the most eco-healthful manner as a single, living organism, sounded too mythical. Three decades on, the hypothesis is now mainstream, giving us a view of life as existing in rhythmic harmony, where population levels rise and fall as waves, maintaining a dynamic balance among species; and where waves of gas and mineral flow back and forth, controlled by the biosphere. All motion, all rhythm. The late Lewis Thomas, in his book *Lives of a Cell*, speaks of viewing planet Earth from space: "If you could look long enough, you would see the swirling of the great drifts of white cloud, covering and uncovering the half-hidden masses of land. If you had been looking for a very long, geologic time, you could have seen the continents themselves in motion, drifting apart on their crustal plates, held afloat by the fire beneath. It has the organized, self-contained look of a live creature full of information, marvelously skilled in handling the sun." That's Gaia. That's life. That's rhythm in nature, on a scale we cannot easily perceive down here on the planet's surface.

Even when the scale is smaller—say, a few thousand square miles—our perceptions of the dynamics of life are often myopic because we lack the dimension of time. Here's a story of an ecosystem in southern Africa, a place managed by park personnel who had a view of how the park should look to please the many tourists. It is a cautionary tale whose message is twofold: first, the rhythms of nature are pervasive and powerful; and second, these same rhythms make life complicated.

Chobe National Park in northern Botswana is typical of many ecosystems in southern and eastern Africa. There are many large herbivores, some of them migratory, including giraffes, buffalo, elephants, zebras, wildebeests, and impalas. Lions, hyenas, wild dogs, and jackals thrive here, too. A mosaic habitat of grassland and acacia woodland harbors a rich array of bird and insect species. Altogether, the park offers an abundant diversity of species of the sort that people think of when they hear the word *wildlife*. Park managers would like to maintain this diversity because it is attractive to tourists and because it is perceived of as being the way it should be. They are, however, facing a severe challenge: the acacia woodlands are being destroyed, principally by elephants, and no new trees are growing. If the woodlands shrink to mere remnants of their present selves, the managers believe they will have failed because they want to keep things as they are. To do so would, however, not only be wrong ecologically but also impossible. A look at the ecological history of the park reveals why.

The Savuti Channel is the major source of surface water in the area. When full, it flows from Angola via the Linyanti swamps and empties into the Savuti Marsh (which currently is grassland). It was full in the late 1800s, dried up around the turn of the century, and remained dry until the mid-1950s; it then refilled. In 1982 it dried up again and remains so. Soon after the channel dried up in the early years of the 20th century, a massive outbreak of rinderpest occurred in the area.

Those two events played midwife to the current acacia woodlands: lack of water encouraged the elephants to seek water elsewhere (hunting also reduced their numbers); and the rinderpest epidemic devastated the ungulate population, such as buffalo, zebras, and antelopes. As a result, browsing pressure in the area was suddenly very light, allowing acacia seedlings (a favorite food of many browsers) to mature into trees. By the time the elephants and ungulates returned, extensive acacia woodlands had been established. "What we observe today, the coexistence of lots of elephants and extensive acacia woodlands, represents a very narrow window in time and is apparently not sustainable," notes Brian Walker, an Australian ecologist who has made a detailed study of the region.

It is not sustainable because as long as there are healthy populations of elephants and ungulates in the area, no acacia seedlings will survive to maturity. The animals would have to be removed if the woodlands were to thrive again. "The question," states Walker, "is whether managers and tourists are prepared to accept a 10-to-15-year period with virtually no animals to see."

Probably not. The current species diversity of the park is natural, of course, but it is generated by waves of substantial environmental change that take place over many decades. And change is what park managers often resist (at least when they see something of value apparently disappearing).

Ecosystems are in a constant state of motion, both in space and time, and at any point some species' populations will be in decline while others may be booming. Constant change is vital as an engine of species diversity. "Conservationists should spend less time worrying about the persistence of particular plant or animal species," warns Walker, "and begin instead to think about maintaining the nature and diversity of ecosystem processes." Ecosystems on any scale, down to the pond in the meadow, are in a state of constant change—wavelike rhythms among and between all the species that constitute it. These waves are the source of ecosystem health.

A similar illustrative ecological tale emerged in the United States in the summer of 1988. Yellowstone National Park was a scene of devastation as fire raged unchecked and uncheckable through almost a million acres of forest and shrubland, ignited by a single lightning strike on June 22. Despite the largest firefighting effort ever in national-park history, the fires raged until October, when they were finally extinguished by early autumn rains.

There was much public anguish about what had caused such a conflagration. Pundits asked, "Why had fire-management practices failed so badly?" The answer, as it turned out, was that such practices had not failed. Rather, a fire of this magnitude was to be expected at some point, because waves of ecological succession had essentially turned the park into a giant tinderbox.

Forest fires spread rapidly when flammable material, such as dead trees and branches, accumulate in the underbrush. During the early part of the 20th century, forest managers' response to a fire was simple: put it out. They then realized that this practice was making the forests more vulnerable to devastating fires, as tinder accumulated. As a result, forest managers backed off and from 1972 on allowed naturally set fires, often from lightning strikes, to burn; they even had a program of controlled burning to reduce the dangerous underbrush. Occasional modest fires were said to be necessary to the ecological health of the park. The gargantuan 1988 fire therefore came as a shock and a source of self-doubt: "What were we doing wrong?"

Nothing, as it happened, and the answer was to be found in a proper ecological, historical perspective. A fire of similar proportions had occurred almost three centuries earlier, in the 1700s. Vast swaths of old-growth lodgepole pines, the park's dominant tree species, were wiped out. That set the scene for the first stage in "ecological succession," the phrase biologists use to describe the sequence in which different species establish themselves as an ecosystem matures. Succession typically goes through several stages, or waves.

In the first stage of the Yellowstone ecosystem, small plants and lodgepole pines spring up among the dead trees left by the fire. In fact, lodgepole pines depend on fire for their existence; their seeds cannot germinate unless they are exposed to fire. This first stage lasts about 50 years, during which time the forests are not very flammable. In the second stage, which lasts about 100 years, the pines form dense stands up to 50 feet tall, and their shade blocks the growth of vegetation below them. The forest continues to be not very flammable. The stands of pine begin to thin out in stage three, and ground vegetation gets a strong foothold. In the last stage, which lasts for 50 or more years, the pines begin to die off as they reach their limit of longevity. The forest is now primed for a major fire, to be fueled by hundreds of thousands of dead and dying pines. Once the right conditions are in place, with a hot, dry summer and strong winds, as happened in 1988, the conflagration is unavoidable.

Many observers in the media and in politics viewed the fire as the disaster it appeared to be: the media showed hundreds of thousands of acres of blackened landscape. An editorial in the Richmond *News Leader* stated, "If you want to see the world's largest charcoal grill, just visit Yellowstone. Be sure to say, 'Thank you environmentalists!'" This paper and others failed to recognize the rhythms of nature at work in the park. The 300-year cycle of birth, maturity, and death is fundamental to the very existence of the forest and its ecological richness.

The spring of 1989 bore witness to the beginnings of a new cycle, as lodgepole seedlings sprouted and began growing an inch or two each year. Wildflowers were resplendent, and grasses and shrubs painted a rich green canvas, proclaiming new life. Yellowstone was not dead; it was at the end point of one cycle and the onset of the next, a three-century rhythm that has gone on for millennia.

Nature is rhythms at all scales, and if we don't understand that, we fail to understand the very nature of nature. Cycles of life and death are the fount of creativity, and if we try to intervene, controlling nature and its ecosystems in the form we think it *should* be, we stifle nature.

When we think about species in ecosystems, in the context of evolution, the phrase "survival of the fittest" often crops up. And it is true that, through the process of natural selection, species do become supremely adapted to their environments in a kaleidoscope of ingenious ways. But the history of life is not just the history of continuous adaptation, leading to the expansion of the diversity of organisms on Earth. It is also the history of extinction events, some of them catastrophic, such as the spectacular collapse of species numbers 65 million years ago, when the dinosaurs disappeared. There have been five such mass extinctions in Earth history, each event changing the face of the planet, literally. Once-dominant groups of species vanish, and new ones take their place. Mammals had been an insignificant presence for millions of years during the reign of the dinosaurs, for instance, but it wasn't until the demise of the Terrible Lizards that mammals came to the fore and took a leading role. As with all major extinctions, a wave of new life followed a wave of death.

What many people are unaware of, however, is that major die-offs are not restricted to the Big Five, as the most famous mass extinctions are known. In recent years, paleontologists have discovered that they occur with eerie regularity, every 29 million years or so. There is much discussion about what drives this periodic culling of species, but the best guess is impact from asteroids. Good arguments can be made for how asteroid impact could occur on a regular basis. But the point here is that Earth history is punctuated regularly with substantial extinctions, and each event is followed by the rapid blossoming of new species, a repeated resetting of the stage of life. Extinction was for a long time largely ignored as a factor in shaping Earth history. But the fact is that we humans are here today, in the company of the 50 million or so species around the globe, largely because of the waves of extinctions and rebounds. It's a wave of life and death on a grand scale.

It was three in the morning one April day in 1971 in an operating room at Montefiore Hospital. Irv Dardik and a surgical resident had been called to do emergency surgery on an elderly woman who'd had a blood vessel graft in her leg a year earlier because her own vessels had become blocked.

Now the graft itself was blocked and needed to be cleaned up.

Blocked or partially blocked blood vessels in the legs are relatively common in elderly people as a result of various diseases, such as diabetes and atherosclerosis. Walking can become difficult and often so painful that the individual has to sit after just a few steps. Eventually, the nerves and muscles in the leg become starved of nutrients and oxygen because of the restricted blood flow, and the tissues become gangrenous, or, quite simply, begin to rot. The only way to avoid the grisly prospect of amputation is to replace the blocked blood vessel with a substitute graft.

"As I was approaching the graft, I was telling this guy about the problems associated with these kinds of grafts," Dardik recalls. "Some of the grafts are synthetic, made from Dacron, for example, and they are often not very satisfactory because they can cause clotting and encourage infection. The one in this woman was a blood vessel from a cow, and these aren't great, either. They often get rejected. That's what was happening in this case." As he was cutting through a tremendous amount of scarred and inflamed tissue, Dardik mused on the perfect blood vessel graft: "It should be resistant to clotting and infection and be flexible, strong, and readily available. No such animal exists."

When he finally got to the troubled graft, Dardik added one more desired property: "It should be human in origin," he said. It's true that in some cases a blocked artery can be replaced by taking the saphenous vein from the patient's own leg and use that as a replacement. But there were problems here, too. For instance, they aren't always available, because they are often diseased or may have been removed because they have become varicose. In any case, the operation imposes extra surgical stress on patients who are often in pretty poor medical condition. Finding a better alternative could revolutionize vascular surgery.

Dardik opened up the diseased graft, and in one of those glorious moments of instant insight, he realized what a better alternative could be. Blood had clotted in the graft, but not in the usual way, which would be a single, amorphous wad of coagulated blood. "The blood had separated into a clump of red blood cells and gelatinous plasma," recalls Dardik. "The woman had been on anticoagulants. Perhaps she had been lying down for long periods in her nursing home. In any case, the blood had somehow separated. When I scooped out the jellied plasma I turned to the resident and said, 'Remind you of anything?' The guy shook his head. 'Well, it reminds me of Wharton's jelly.'" Wharton's jelly is the gelatinous stuff that surrounds the blood vessels in umbilical cords, and it oozes out when the cord is cut in the birthing room.

The recognition instantly took Dardik back to a night a year earlier when he was on obstetric duty at Morrisania Hospital in the South Bronx. "A woman comes in and she's yelling and screaming," he recalls. "I was sleeping, and the nurse wakes me up and says, 'You've got to come and examine her.' She wasn't dilated, so I went back to the doctor's room. Several times the nurse dragged me into to examine the women, but nothing had changed. I went back to my room again each time. This was going to be the woman's 11th child, so I knew the birth could happen quickly. It did. I hear her screaming again, so I saunter over to the delivery room, and there she is, standing on the bed—the baby has delivered and is bouncing on the end of the umbilical cord. What a sight that was!"

Remembering that sight, Dardik said to the Montefiore resident, "That's it. A blood vessel from a human umbilical cord gives us all the properties we need in a perfect graft. It's a human, it's flexible and strong, there's tons of it readily available, and no one has ever used it as bypass material." Dardik finished the case as quickly as he could and, even though it was four in the morning, called his brother Herbert. "I said, 'I have the answer to the bypass problem!'" recalls Dardik. "He thought I was nuts. He said, 'Umbilical cords from babies? Are you crazy?' I said, 'Maybe I am, but I know I'm right!'"

There are two arteries and one vein in umbilical cords. Dardik saw that the vein was the ideal size for what he had in mind. He went to bed, excited and looking forward to sharing his insight with the higher-ups at Montefiore. They would see what a great idea it was, he thought, and would do everything to help make it a reality, right?

Dardik should have listened to his brother. "I am always skeptical," says Herbert. "My position is, 'OK, you have an idea. Now you have to prove that it's right.' But Irving has no skepticism when he has an idea. He has an idea and right away he knows he's right. He gets angry with me sometimes over this kind of thing. When an idea of his works out he comes back to me and says, 'See, you didn't believe me. I was right!' It's not that I don't believe him. I just need an idea to be proved before I will believe it. In science and medicine, healthy skepticism is a good thing." As it turned out, the Dardik brothers were to face more than just healthy skepticism.

Within a few days of his insight, Dardik went to see Marvin Gliebman, chief of surgery at Montefiore, and told him about his idea. "His immediate response was, 'What are you talking about? The umbilical cord is programmed to die after nine months.' I wasn't too surprised that he was negative. There's often tension between full-time hospital staff and people like me. In any case, I persisted, saying I thought it could revolutionize vascular surgery." Gliebman called in Frank Veith, who was head of vascular surgery and involved with lung transplantation. "I went over my ideas with Veith, and his reaction was the same as Gliebman's. All I wanted was some laboratory space to research the idea, but, hey, they're not going to give a clinical surgeon who's making a lot of money lab space in a hospital."

Vascular surgery was in its infancy as a surgical specialty in the early 1970s: no journals dedicated to it, no professional society. Dardik's brother Herbert was one of the founders of the specialty. Dardik himself hadn't planned to be a vascular surgeon; it was his experience as a surgical resident that took him there. "At Morrisania, there were a lot of trauma cases," he explains, "people being shot and stabbed, blood vessels torn apart. So I saw a lot of vascular surgery. Often I had to do things in the ER for which there were no standardized procedures, to hold on for dear life. It was fascinating, and I liked that. I liked the technical aspects of the surgery. It's very artistic. And I found I was good at it. Because it was a young specialty, there was a lot of room for experimentation, rather than simply following established rules. The umbilical cord story is a good example."

Over the next few months, Dardik turned his basement into a makeshift laboratory, where he experimented with different chemical treatments of the umbilical vein. It looked and smelled like a chemistry lab, except there were jars containing umbilical cords in different fluids, which gave the place a macabre ambience. Meanwhile, Herbert was planning experiments of his own. "I got some umbilical cords from Morrisania, took them to my house in Teaneck, and pickled them in formaldehyde, then forgot about them," he recalls. "Spring and summer passed, and finally my wife, who does laundry in the basement, gave me an ultimatum: 'Either get rid of these things, or do something with them,' she said. I was going to throw them out, but then I thought that would be a waste. I got a table out in the backyard, got a dissecting kit, and started to work on them. My son, who was about seven at the time, was sitting right there on the table, watching me. I can't imagine what he thought, but he was fascinated. The cords looked awful, wrinkled and discolored. I needed new ones. I got in my car and drove over to Irving's house in Tenafly, and we talked about what he had been doing and what we needed to do together to get this thing tested."

One key test was to see if the vein would hold up physically once it was removed from the umbilical vein. "We 'borrowed' a miniature heart pump from a lab at Montefiore," Herbert recalls. "One of the guards stopped us as we were leaving the building, because we really shouldn't be taking the pump off the hospital premises. But he knew us and turned a blind eye." In the best tradition of pioneering experimenters, Dardik drained a pint of his own blood and decanted it into the pump system, to which the umbilical vein had been attached. "We let it run over the weekend," says Herbert, "and we were delighted to find that, not only did the vein stand up, but also the blood didn't clot."

In a moment of zany creativity, the brothers fantasized about establishing a nationwide BBB network, or Belly Button Bank network. "The idea was that when someone was born, their own umbilical cord would be saved," explains Dardik. "Then, 75 years later, when that person needed a vein graft, the cord would be taken out of deep freeze and the vein would be taken out for the necessary transplant. Brilliant! But then we got down to serious work."

Dardik needed a lab where he could do experimental grafts with animals. With the labs at Montefiore closed to him, he had to find an alternative. Dardik had heard of New York University's Laboratory for Experimental Medicine and Surgery in Primates (LEMSIP), located in tony Tuxedo Park, 25 miles northwest of New York City. Interestingly, the lab is just a stone's throw from a secret laboratory where British and American scientists worked to thwart the Nazi menace in Europe in the early 1940s. That lab was buried in the bowels of the fabulous mansion of Wall Street tycoon Alfred Lee Loomis. LEMSIP's accommodations were a tad more modest, but its research reputation in the field that interested Dardik was towering.

Dardik called the lab's director, Jan Jankowski, and arranged a meeting. Herbert was skeptical that Jankowski would agree to let them work on baboons, which are a whole lot more problematic than, say, laboratory rats. But Dardik was resolute. He was determined to have baboons because, as they are primates, they are biologically closer to humans than are most other experimental animals. They are, of course, bigger than rats, and that was important in the work Dardik and his brother had in mind. "I walked into Jankowski's office," says Dardik, "and told him that I had this idea that is going to change medicine and surgery. I asked, 'How do I get baboons?' Jankowski looked at me and said, 'Dr. Dardik, I know you are from Montefiore, and I know a little bit about you from your research papers. But do you understand that when you bring baboons into the lab you have to quarantine them for six months, and they are expensive? You can't just say, "I want to take some baboons, and then do some surgery." It doesn't work that way. You need a research protocol, and it is going to take a year or two just to get started.'"

It looked as if Herbert was right. Undaunted, Dardik replied, "I'll make you a deal." In his typically enthusiastic manner, he explained: "I think that what I am proposing is so exciting that you will want to give me *your* baboons." He went on to describe his insight about using veins from umbilical cords and the tremendous impact his proposed procedure could have on blood vessel transplants. He also said that the work would produce interesting results in the immunology of cross-species transplantation, Jankowski's field of research interest.

Nobody at Montefiore knew what Dardik and his brother were up to. Dardik would continue to do surgery at various hospitals in the New York area during the day, then secretly go to LEMSIP to do experimental surgery at night. LEMSIP was his new cave, where he could safely do the work he wanted to do, out of sight of the disapproving establishment. Herbert usually joined him.

The initial results were disappointing. Some of the vessels burst in places, and often there was immunological rejection. Nevertheless, this preliminary work showed that the procedure was technically feasible, work they reported in the *Journal of Medical Primatology* in 1973.

The Dardik brothers concluded that they couldn't use the umbilical vein "as is." It needed to be toughened. Over a period of 18 months, they experimented with various hardening agents in Dardik's basement. Finally, they settled on glutaraldehyde, which is used for tanning leather. It worked brilliantly. Not only did the chemical make the blood vessel more durable, so that it no longer burst, but it also produced changes on the surface of the vein that prevented immunological rejection—a tremendous bonus.

After a series of successful transplants in baboons, Dardik and his brother turned to human experimental subjects. "What we could do back then, we can't even consider doing today," explains Dardik. "My brother and I were on the staff of Union Hospital in the Bronx, a tough area, a little hospital with about 50 beds. I could try new techniques without any red tape. You tell someone, 'You've got gangrene; I'm going to save your leg.' Nobody cared. You have patients who are going to have a limb amputation, and you offer them a way of saving it, and they say, 'Go ahead, try it.' We began to get wonderful results, which got better and better as we modified the procedure with experience."

By early 1975, it was time to report the work in a scientific paper, a job that fell to Herbert, given Dardik's famously chronic problems with writing. The paper was titled "The Use of Human Umbilical Cord as a Vascular Prosthesis," which they planned to submit to a journal called *Surgery, Gynecology, and Obstetrics*.

As with all scientific papers, there is etiquette to follow, which Dardik and his brother discussed at length. The work had been done at LEMSIP, so that had to be acknowledged, of course. The issue was whether Montefiore Hospital's name should also be on the paper, as the two brothers had surgical posts there. "I felt obligated to tell the people at

Montefiore," explains Dardik, "and I thought they would take some pride in having the hospital's name on the paper." Herbert agreed. "So Herbert and I went to see Marvin Gliebman, to show him a copy of the paper we were about to submit. He agreed that Montifore's name should be on it."

What happened next was not at all what the brothers had expected. Gliebman said, "And you will have to have Frank Veith's name on it, too. He is chief of vascular surgery here, and he is entitled to have his name on the paper. You need to put Frank's grant number on the paper and acknowledge that we supported you."

True, the head of a lab will often be named on a paper about work done in the lab, even if he or she hasn't actively participated, questionable though that practice may be. But here, the situation was quite different. Dardik told Gliebman, "Wait a minute. Veith wasn't involved at all, no input. The work wasn't done here; it was done at the NYU lab. And you even tried to persuade us not to do the work at all." Dardik and his brother left, fuming. "We had a lot of pressure from other surgeons at Montefiore to conform with what Gliebman was asking for," says Dardik, "and the issue simmered for weeks. Finally, we decided to send the paper in, with just NYU's name on it, excluding any mention of Montefiore. No one at Montefiore knew we had submitted the paper. They thought we were just sitting on it."

The paper was published a few months later in *SG&O* as the lead article, a clear recognition of the importance of the work. The brothers were excited and proud and looked forward to helping scores of people who might otherwise lose limbs. "The next thing you know," says Dardik, "my brother and I were summoned to meet with the head of the hospital." This was in April 1975. "Veith and Gliebman were there, and a lawyer, too. The head of the hospital said, 'You had no right to have published this paper. Dr. Veith has been doing research using umbilical cord blood vessels as a vascular prosthesis in lung transplantations. And the idea to use umbilical cord blood vessels in this way isn't yours. It was developed by Dr. Levovitz at Brooklyn Jewish.'"

A little more than a year earlier, Dardik's brother-in-law had visited Dardik's Tenafly house and had seen Dardik working with umbilical cord vessels. "I was a little suspicious of his interest," Dardik admits, "so I didn't let him know that we were using glutaraldehyde to treat the vein. I told him we were using dialdehyde." At the time, his brother-in-law was a surgical resident at Brooklyn Jewish Hospital. During a presentation of a patient's case, Levovitz explained that the patient needed a vascular bypass and asked for suggestions about how they might proceed. Knowing what the Dardik brothers were doing, his brother-in-law suggested using the umbilical vein.

Levovitz thought it was a brilliant suggestion. Levovitz had his colleague Bruce Mindisch pursue the idea, as well as the use of the cord in dialysis patients. "We didn't know about this at the time," remembers Herbert. "But later I saw a notice of a meeting of the American Society of Artificial Internal Organs. On the speakers' program was Mindisch, who was going to present his work. I contacted the meeting chairman and explained the situation. He told me that he couldn't withdraw the paper, but he would allow me to present a discussion of our umbilical vein work. I did that and made it very clear whose idea it was. This was before our paper had been published."

As for Veith's use of umbilical vessels in his own work, the story is a little more complicated. At the time, he was applying to the National Institutes of Health for a multimillion-dollar grant. In the application, he described his plans to use the vein as a transplant and said that the work by the Dardik brothers was done under his supervision. "This was news to us," says Dardik. "Our only contact with Veith during the work was tenuous, at best, and involved our use of his laboratory technician at LEMSIP for one evening session only. Doing the work in the evenings, as we had been doing, was tough, and I thought it would ease the burden if we had some help. Herbert was uncomfortable about including this guy, and so he didn't come again."

The overseers at NIH were intrigued by Veith's grant request, especially the use of umbilical vessels for transplants, and asked for documentation in the form of a publication. Herbert remembers, "The next thing I know, there are letters going back and forth between the NIH and Montefiore lawyers, trying to sort all this out. In the end Veith had to back down, but he got his grant anyway. We won the battle, but we lost the war because after that we were personae non gratae at

Montefiore. We were denied access to patients, denied access to operating facilities."

"What saddens me about this episode," says Dardik, "is that it was the first time in my experience that people in the medical profession weren't focusing on patients' well-being. Instead it was all about ownership of the idea and who was going to benefit from it. And here we were, after having been told initially that our idea was ridiculous, that it wouldn't work, seeing other people claiming it as their own. But we weren't worried because we had proof of where the idea came from."

What the Montefiore brass didn't know was that, back in 1971, when Dardik first had the idea for the work, he had had a crucial conversation with George Robinson, chief of cardiac surgery at Montefiore. Robinson told Dardik, "Be smart, Irv—apply for a patent. It will give you protection. You never know what might happen. The fact that you get a patent says it's your idea."

Dardik followed that advice and got a patent within months. "So here I am, sitting in the office of the chief of the hospital, hearing all these claims," says Dardik. "I said, 'Excuse me, I live across from the hospital, and I have something there I think you should see.'"

He left and came back within minutes with the patent in a briefcase. "I walked in, sat down, and said, 'You say this is your idea. What is your proof?' Veith replied, 'The fact that I applied for a grant to do the work. Here it is.' I took out the patent, put it on the table, and said, 'Here is the patent that Herbert and I got approved, and it is dated long before your grant proposal.' They were in shock. They just sat there."

Dardik and his brother contemplated taking legal action but decided against it. Instead Dardik opted to leave Montefiore and take up a post as staff surgeon at Englewood Hospital in New Jersey.

The two brothers pressed on with their pioneering procedure, which began to get noticed by the national press, such as *U.S. News and World Report* and *Newsweek*. By the fall of 1977, Dardik and his brother had operated on almost 150 patients, mostly elderly, suffering from diseases that disrupted blood flow, such as diabetes and heart disease. All had been in danger of losing a leg. The new biograft procedure saved more than 80 percent of the patients from that fate. *Parade* magazine took notice and on August 7, 1977, published "Umbilical Cords Provide Grafts for Arteries."

The opening paragraph of the article read as follows: "In a New Jersey hospital a few months ago, a 26-year-old man faced the loss of a leg after an accident that had grossly damaged a major leg artery. Today he has full use of his leg. And for the first time in years, a 65-year-old man and a 72-year-old woman are free of crippling from blockage of leg arteries so severe that he could take only half a dozen steps without agonizing pain; she suffered pain even at rest, and both were threatened with gangrene." The magazine went on to sketch out the umbilical cord story and explain why it offered so much promise.

To thousands of people across the country, that story offered an unexpected beacon of hope in what seemed like an inevitable path to the loss of a limb. For instance, on that same Sunday, John Rosza, a 50-year-old industrial worker, was in a Connecticut hospital. The toes and ankle on one of his feet were black and rapidly becoming gangrenous. Circulation in the leg was blocked, and there was no pulse. The leg was scheduled for amputation within 48 hours. Rosza's sister-in-law read the *Parade* story and hurried with it to the hospital. When Rosza saw it, he called his surgeon and cancelled the amputation, went to Englewood Hospital, and persuaded the Dardik brothers to perform a biograft operation instead. Within two weeks the pain in Rosza's leg was gone and the gangrenous areas were healing. Rosza called it "a miracle."

Also on that Sunday, Audrey Servello of Fresno, California, woke to find her husband by the side of her bed, holding the issue of *Parade* and saying, "You have to read this." Like Rosza, she was due for a leg amputation because the toes of one leg were gangrenous; the other had already been removed. Losing the leg was the only recourse, her surgeon had told her. Servello and her husband flew across country, went to Englewood, and were accepted as the Dardiks' patient. Within days there was a pulse in her leg; it felt warm for the first time in years, and she was eager to be up and about. It was, she said, "an answer to my prayers. I had been left without hope."

Rosza and Servello were just two of hundreds of people who had been without hope, saw the magazine article, and made the successful pilgrimage to Englewood, all of which was reported in a November issue of

Parade that year. The Dardik brothers and their colleague, Dr. Ibrahim Ibrahim, were swamped with patients pleading for the new procedure and were working extended hours. More physical therapy staff were hired to cope with the burgeoning workload. "It was a heady time," remembers Dardik. "It wasn't that we were working like crazy, doing this technically beautiful procedure. It was the fact that we were giving people their lives back at a point when they thought there was no hope. As a doctor, there's nothing that can beat that."

Within a year, scores of surgical teams around the world were using the umbilical cord biograft, particularly in lower-leg vascular disease. The popularity of the technique soon expanded so much that Dardik and his brother were enjoying a healthy cash flow in royalties as a result of the 1971 patent, which was good for 15 years. Monetary rewards aside, Dardik and his brother also received professional recognition when the American Medical Association bestowed on them its highest honor for innovative research and practical application, the prestigious Hektoen Gold Medal.

6

View from Olympia

"How can the events in space and time which take place within the spatial boundary of a living organism be accounted for by physics and chemistry?"

—ERWIN SCHRODINGER

BEFORE HUMANS BEGAN TO HAVE AN IMPACT on the environment, Earth's biological and physical systems interacted in a coherent and sustainable way, interrupted now and then by major extinctions from which there was rapid recovery. This coherence and order is the outcome of communication through ever more complex hierarchies, mediated by superesonant wavenergies, or SuperWaves, at each level. That coherence and order is now being disrupted and at an accelerating rate, as humans exploit and destroy the environmental and physical resources of the planet. We have created for ourselves an artificial environment in terms of where and how we live and work, severing our connections with nature. The coherence of the ecosystem we inhabit, the hierarchy of wavenergies that maintain the health of the whole, has been disrupted. These disruptions, Dardik argues in his theory, threaten the future health of both the global ecosystem and ourselves. Indeed, they already have.

Anyone who cares to be aware of what is happening in the world knows only too well that the environment is under assault: rain forests, the lungs of our planet, are being felled at an alarming rate; the list of newly extinct and endangered species grows daily; and greenhouse gases are building to a point where the Earth's atmosphere is not only warming, but weather patterns are changing and displaying violent variations. Beyond the immediate impacts of these effects, however, lurks a greater danger. The biological and physical systems of our planet are complex and interconnected, honed by millions of years of evolution and coevolution. Such systems are very robust and can sustain a lot of damage without global consequences. From a SuperWave perspective, however, comes the recognition that there will be a point at which one more small increment of disruption will lead to global ecological collapse.

Think of an airplane flying in the sky. How many rivets can be removed and the plane still remain aloft? Quite a few, probably. But there would come a point when the plane falls apart and plunges to Earth if one more rivet is removed. It's not the best analogy, admittedly, because a plane is a machine; complicated, it's true, but not complex in the way that ecosystems are complex. There can come a point at which the loss of just one more species can cause ecosystem collapse through unpredictable interactions among those remaining. Many biologists are already talking about a man-made, sixth extinction being quietly under way. (The previous five mass extinctions were from natural causes.) How many more individual species' extinctions are necessary before the coherence afforded by SuperWave interactions unravels and the sixth extinction is no longer a quiet threat but a loud reality?

Meanwhile, Dardik argues in his theory, humans are already suffering the effects of our disrupted connections with nature that result from living in a technological, civilized society. The disengagement of SuperWave connections between humans and the environment that once nurtured physiological function and health in humans has led to linearized, less complex wavenergies in each of us, he says. Our waves have become flattened.

Less-than-optimal wavenergies lead to less-than-optimum physiological function—and disease. The current epidemic of chronic diseases in Western societies, which are virtually absent in peoples living in harmony with nature, is the inevitable outcome. However, armed with an understanding of chronic disease through the SuperWave Principle, we can counteract it, says Dardik. The effect of the cyclic exercise protocol that is Cardiocybernetics is to restore the complexity and robustness of an individual's wavenergy, thus reversing disease. Similarly, the health of ecosystems can be restored by stopping, and reversing, environmental destruction and, for example, establishing physical connections between national parks, an idea that has been discussed by some ecologists.

The SuperWave Principle, overall, is not a theory of relationships or a mathematical formulation of a model, but a simple understanding, born out of life, that recognizes the inherent continuity of the natural universe. The principle is the physical reality as well as the patterns of order of the natural universe. As the physicist Heinz Pagels wrote, "Part of the answer, I believe, will turn out to be that we are asking the wrong question, making a false distinction between the transcendental and the natural world. But to see that that is the answer will be quite an accomplishment, one that will change our very civilization."

Despite the emotional and financial rewards that Dardik was reaping from his surgical practice, he was becoming restive. The politics of the medical world irked him, and he was beginning to feel caged again. More and more he turned to another passion in his life: sports. By the late 1970s he was deeply involved as the physician to U.S. national sports teams, including Olympic teams. Life was frenetic, with half his time spent doing vascular surgery and half working with athletes, finding ways to improve performance at Olympic sports medicine centers around the country. "My partners in our surgical practice, including my brother, weren't especially thrilled that I spent so much time away, while having my share of the practice's income," he recalls. "But being with the Olympics was like breathing fresh air for me. It was such a relief."

The shift toward sports involvement began one August day in 1969, during the eighth staging of the Maccabiah Games, the so-called Jewish Olympics, held in Tel Aviv, Israel, every four years. First held in 1932, they are named after the Jewish warrior Judah Maccabe, who defeated the Greek king Antiochus, retook the holy city of Jerusalem, and established Hanukkah in 165 BCE to celebrate the victory. Dardik had competed in the 1957 games while an undergraduate at the University of Pennsylvania, garnering two medals, a silver and a gold, in sprint and relay races. His dual passion for Israel and sports led him to continue attending the games after that, acting as a doctor to the U.S. team. He also became team physician at the 1971 Hapoel Games, also held in Tel Aviv.

On the eve of the 4 × 400-meter relay race in the 1969 Maccabiah Games, one member of the U.S. team pulled a muscle and had to withdraw. It was little short of a disaster because the team was widely expected to win the gold medal. Bob Giegengak, the Yale track coach, and Irv Kintisch, the Columbia track coach, the coaches for the U.S. track team, were sitting around, trying to figure out what to do. Haskell Cohen, a U.S. official with the games, was listening to their conversation and said, more as a joke than anything else, "Why don't you let Doc Dardik run the opening leg? It will make an excellent story!" The coaches laughed, then continued with the serious issue at hand.

"After a while, Kintisch came to me and said, 'You know, that idea of yours isn't as nutty as it sounds,'" Cohen later recalled. "Irv has been working out constantly and is in great shape for a 32-year-old. We're running him tomorrow." Dardik was in fact more involved with the athletes than the average team physician: he loved running, so he worked out with them, running intervals and so on and doing quite well for a man of his age who was just having fun.

Even though Dardik hadn't run competitively in years, he was thrilled to have the opportunity to do so now. "They told me that I was to run the opening leg, and that if I could just keep within 10 yards of the other teams, they had three guys who would be able to make up the deficit," says Dardik. He did better than that, finishing his leg in 49.1 seconds—two yards ahead of the closest rival team—and the gold medal was won. Years later Kintisch sent Dardik the baton, inscribed with the relay team's signatures. He wrote, "Dear Irv, I just came across this baton, and thought you might like to add it to your collection of memorabilia. A fine relay time and a gold medal, 1969."

People in the sports world heard what happened in Tel Aviv that year, and Dardik was soon being invited to be the physician for U.S. teams: first locally, at the Millrose Games in New York and at New York Athletic Club meets; then nationally and internationally, beginning with the Pan American games in Mexico in 1975, then at the winter and summer Olympic Games in1976, 1980, and 1984. "I hadn't fulfilled my dream of competing in the Olympics," Dardik says, "but being a team physician, I got to be part of the whole occasion, marching into the Olympic stadium with the athletes. What a thrill that was. You are standing with your team, in uniform, your country's flags waving. Then you march in, and there are a hundred thousand people in the stadium; your country's anthem is playing, and you know you are one of 130 countries there. It's unforgettable."

Dardik's lifelong passion for running was only one reason he reveled in being team physician. He's not just a sports jock. As he typically does, he took the involvement with athletes, and with his patients, as an opportunity to learn, to make sense of life and of his own life. A clue to his childhood conviction that he had important work to do was to be found in here somewhere, he believed. Finding the path to that grail would determine where his professional life would take him—he knew that.

On one hand, he was dealing with athletes who were healthy, and on the other, he was treating patients who were sick. But he saw that they were not so different. "It intrigued me that the athletes and the cardiovascular patients were all in the same boat, mentally, physically, and nutritionally," he explains, "in terms of trying to improve and optimize performance and to prevent injuries and the recurrence of diseases. The same processes are involved. It is the same physiology that enables us to survive and for athletes to get fit, as it is in people who get sick. So it was a question of, what is the nature of that continuum? How are they linked?"

In 1975 he acted on this intuition of the existence of a fundamental link when he established what he called the American Sports Medicine Training Center in a huge rented indoor space in Englewood, New Jersey. It was just down the main street from Englewood Hospital, so Dardik could go back and forth when necessary. "I wanted to find out the impact of exercise on health," says Dardik. "My focus was primarily on juvenile diabetes sufferers because at the time, no one had looked into the effect of exercise on the symptoms of the disease, or any disease for that matter."

Dardik recruited partners in the project, most notably Bill Simon, secretary of the treasury under presidents Nixon and Ford. Simon was also involved with the Olympics and was intrigued with notion of the impact of exercise and health, so he enthusiastically joined Dardik in figuring out how to proceed with the venture. He also persuaded his friend Bill Casey, who would later head the CIA under President Reagan, to be part of it. The funding for the project, however, came from Dardik's personal purse, which was amply swelled by fees from his practice and later by royalties from the biograft. "I wanted to inspire the kids involved in the diabetes project, so I brought in Olympic athletes," says Dardik. "Jimmy Dietz, the rower, was there, as was Fred Samara, currently the track coach at Princeton. In addition, we had two Olympic fencers, a kayak star, and a medalist in judo—eleven Olympians in all."

The 50 children in the program ranged in age from six years through high school, and they pursued whatever sport most interested them. An exercise physiologist monitored the kids' oxygen consumption and other athletic measures throughout. One long-term goal was to develop the children's cardiovascular condition, to reduce the chances of arteriosclerosis that so bedevils people with diabetes later in life.

The results of the program in the short term? "Fantastic," remembers Dardik. "Not only did the athletes motivate the kids—you could see that in the way they felt about themselves—but they also began to reduce the amount of insulin they needed. I remember that at the beginning, many of the younger kids couldn't bring themselves to do the insulin injections. They relied on their parents to do it. That changed. Most of them began to medicate themselves. They were different people, and that was wonderful to witness."

The center closed in 1977, principally because the financial burden on Dardik—about half a million dollars by this point—became too great. But before it closed, Dardik presented the beneficial results at the National Juvenile Diabetes Foundation's annual dinner. "Lucille Ball introduced me," says Dardik. "I sat next to her at dinner. That was a hoot, but I knew she was very dedicated to the cause of juvenile diabetes. Anyway, I was able to tell them about the impact of exercise on the disease, the first time anyone had done that. It was a lesson I wasn't going to forget, even though I didn't pursue it any further at the time."

But he did continue to pursue sports medicine, exploring ways to enhance athletic performance. Sports medicine was in its infancy at the time, and it was midwifed by the emerging issue of performance-enhancing drugs in sports, such as steroids for building muscle mass and blood doping, in which athletes get a blood transfusion prior to an event to boost their oxygen-carrying capacity. "Athletes were saying to us, 'Why won't the sports medicine community tell us how to beat all those other athletes who are obviously on drugs?'" says Dardik. "They were looking for some kind of magic bullet." No such thing existed, of course. In any case, Dardik was looking elsewhere, to the original spirit of the Olympics. "During the centuries leading up to the ancient Olympic Games," Dardik now explains, "the Greeks evolved an extraordinary idea, rooted in their mythology, that bound 'body, mind and spirit' into an inseparable 'whole personality'—an idea whose force unified ancient Greek religion and culture. It was a powerful idea to me."

It was powerful because it resonated so harmoniously with the young Dardik's love of nature and the wholeness he saw in it; and it resonated with the adult Dardik's quest to treat a patient as an integrated system, a whole person, rather than modern medicine's proclivity to atomize individuals into parts and categories of disease symptoms. "My goal was to find ways of enhancing athletic performance, and enhancing health, through a holistic approach," he explains, "and that was pretty way-out stuff back then."

Dardik's opportunity to pursue this goal fully came in 1977, when Don Miller, president of the U.S. Olympic Committee, invited him to be the founding chairman of the Sports Medicine Council. Following the winter games in Montreal in 1976, in which Dardik was a team physician, concern over drugs in sports was burgeoning. With the Cold War as a backdrop and Soviet bloc athletes winning medals aplenty, often with the suspicion of tainted training, U.S. Olympic officials fretted over bringing coherence to the training of athletes in the country.

The Amateur Athletics Association and the National Collegiate Athletics Association were locked in competition over who would be the supreme governing body for amateur sports. Dardik was asked to meet with the Olympic executive committee for an exchange of ideas about what might be done. Toward the end of the conversation, Dardik argued the need for a national training center, where a diversity of medical and scientific skills could be brought to bear on athletes' needs. The committee agreed. Congress elected to make the U.S. Olympic Committee the governing body for amateur sports, and, as the newly appointed chairman of the U.S. Olympic Sports Medicine Council, Dardik found himself in charge of all of sports medicine in the country, a position he held until 1985.

During his eight years as chairman, Dardik spent an increasing amount of time at the newly established Sports Training Center in Colorado Springs, at other similar training centers around the country, and at the Games themselves around the world. After 1980, he threw himself into the venture pretty much full time after quitting surgery. "It wasn't that I didn't like surgery," Dardik explains. "I did. But I was still suffering from ankylosing spondylitis, and the pain was cycling in and out. Sometimes it was OK, but when you are in pretty much one position for hours on end, as you are doing surgery, I would often come out rigid, barely able to move. It was excruciating. And I was finding the medical environment more and more oppressive, everything scrutinized and done on a mechanistic basis. I had tasted freedom being with the Olympics, and it's hard to taste that freedom and not pursue it fully."

And pursue it fully he did, bringing ideas he was reading about into his sports medicine philosophy. He was reading avidly at the time, in search of the nature of nature, books like *The Dancing Wu Li Masters* and *The Tao of Physics*, glimpses of the weird world of quantum mechanics. "I was reading about the wholeness of the universe, and that everything is connected to everything else," he remembers. "It was about the importance of waves in the quantum world, how waves and rhythms are fundamental to everything, and I wondered whether there was a connection between waves in the quantum world and in human performance. It gave me an approach to take with the athletes."

Dardik brought experts from a wide range of disciplines to Colorado, including conventional exercise physiologists, psychologists, nutritionists, people in biomechanics, and practitioners of alternative medicine, such as herbalists, chiropractors, and acupuncturists. "We were forerunners of alternative medicine in the United States, through the Olympics," claims Dardik. He also recruited Frank Sulzman, director of biomedical research for NASA, and Harvard's Charles Czeisler, both pioneers in chronobiology, the study of rhythms in nature—a field virtually unknown at the time but one that Dardik felt was going to be important, especially in view of what he had been reading on quantum mechanics. For his part, Sulzman wanted to pursue the link between rhythms in nature and in athletes' performance, especially the impact of taking drugs on those rhythms. It was a rich mix of talents that should have been highly creative and productive, but was ultimately frustrating.

"I worked with a world-record high jumper," recalls Dardik. "We were doing biomechanical analysis on how to optimize his jumping action. We told him that if he raised his elbow this way an inch, his knee an inch higher here, changed his angle of approach, and so on, that he should biomechanically jump two inches higher. Instead, he didn't jump as high! The question was, can you develop imaging techniques psychologically, so you can relax the person, in combination with the biomechanics? It was very difficult to put together the pieces—the psychology, the biomechanics. I began to see that the athletes themselves were performing as an inherent continuum of one, a natural flow that is hard to improve piece by piece."

Nevertheless, Dardik persisted and eventually developed a program called the Elite Athlete Project, which got under way in 1980 and continued until 1985, when Dardik broke his association with the Olympics. The project had two goals: first, to investigate the prevalence of drug taking; and second, to find ways of enhancing performance. In addition to specific athletic-skills enhancement and nutritional counseling, the program was strongly rooted in mind-body techniques such as meditation and visualization. But it also included a novel, key element that Dardik developed, and one that would portend his path for the future: Rhythmic Interval Training Exercise, or RITE.

This protocol is similar to regular interval training, which at the time had been around for a couple of decades, and involves bursts of high exertion interspersed with less exertion—sprinting interspersed with jogging, for instance. With the RITE protocol, Dardik had athletes stop or walk slowly between the bursts of high exertion, thus accentuating the wave of exercise and recovery. They wore heart rate monitors to check their cardiac recovery and record specific target heart rates, something that's not done in established interval training.

Two threads wove together here, giving birth to RITE. The first was when Dardik had a conversation with Frank Shorter, the 1972 Olympic marathon gold medalist; the second was Dardik's observations of athletes at the Olympics.

"I met Frank in 1976, and I was asking him about his training methods," recalls Dardik. "He told me that much of what he was doing was interval training, rather than the heavy distance running that others do. He said that he just had a gut feeling that this is what he should do."

The benefits of this approach were reinforced for Dardik when he began to notice that the sprinters at the Olympics were healthier than the marathon runners. "We did a study with exercise physiologists, and we found that some of the marathoners had immune suppression, which made them susceptible to infections—colds and things. Even basketball players or hockey players were much healthier than the endurance athletes. As a sprinter, I sometimes tried to run distances because we were supposed to, that was the way to train; but I didn't feel good. It was a gut feeling that the way to train was through interval training. And my fascination with the emphasis of waves in quantum mechanics encouraged me to enhance the wave pattern, compared with regular interval training."

Curious about why endurance training rather than interval training had become the gold standard for athletes, Dardik sought out two former presidents of the American College of Sports Medicine. "I asked them how they came up with the idea that distance running—aerobics—was the way to go, and why did they not test interval training," explains Dardik. "The reason they gave was extraordinary. They said that when you do oxygen-consumption analysis on a treadmill, with a mask and all the wiring, it was a lot easier to do on continuous run. Why would you want to stop and sit down, with all the wires and tubes getting in the way? They made the assumption that they were focusing on exercise, and of course recovery was not an issue. They were only looking at exertion, so why would you want to do interval training? As a result, all the parameters for what fitness was came to be based on distance running."

The outcome of the Elite Athlete Project was mixed. On the negative side, Dardik and his colleagues discovered that the use of performance-enhancing drugs was more prevalent than had been suspected, particularly in events such as track and weight lifting. On the positive side, the RITE approach seemed to work well for some of the teams. For instance, the U.S. women's volleyball team vaulted into first-place ranking after being dead last, and the national fencers won their first gold medal in the 1983 Pan American Games after being on the program. Cycling and track-and-field athletes began to garner more medals, too. Eventually, more than a dozen national teams took part in the program. "I was very proud of that program," says Dardik. "We really changed something important there."

However, he failed to change sports medicine more broadly, as he had hoped, not just in the attempt to persuade athletes that drugs were unnecessary and unethical, but also in the realm of performance. Although the various experts he had brought together in Colorado Springs provided valuable input in their separate disciplines, they never gelled as a coherent team to produce something entirely novel that could elevate performance. "Every month or two we would get everyone together and have them present their work," explains Dardik. "We wanted them to be able to communicate with one another, to speak the same language. But it didn't happen. We saw that the scientists were very much independent, using their own language, not communicating. They would come in, saying, 'I have the answer.' 'No, I have the answer.' And so on. There was something inherently wrong in the way scientists were trying to communicate, or not. It was very disappointing. Sadly, too many of them simply wanted the kudos of the Olympic rings on their jackets."

Flash back to a mid-June day in 1979, when the phone rang in Dardik's office at Englewood Hospital. Alison Godfrey was the caller, and she introduced herself as the product manager for Medtronic, a company selling various kinds of medical technology. "I have a plan I hope you might be interested in, given your connections with Olympic athletes," Godfrey explained. "Medtronic sells external pain controllers called transcutaneous electrical nerve stimulators; they control pain without the use of drugs, and I thought it would be terrific if our product was used by the Olympic team. Are you interested in pursuing the idea? Why don't you come to Minneapolis so we can talk about it?"

At the time, Dardik's life was frenetic. His involvement with the Olympics was going through a tremendous growth spurt: the Colorado Springs Sports Medicine Training Center had been open just a year; the Elite Athlete Project was gearing up; and he was meeting frequently with national sports governing bodies, both in the United States and Europe—not to mention his continued (albeit decreasing) commitment to his surgical practice. "I have all this stuff going on, and this woman wants me to go to Minneapolis," Dardik remembers thinking. "Is she nuts? I should just politely hang up." But he didn't. Godfrey, then an emerging female titan in the male-dominated culture of corporate America, is a woman of great presence and persuasion. But it was more than her powers of persuasion that stayed Dardik's hand from hanging up. There was an instant interpersonal connection, an alluring chemistry.

The call that in purely practical terms should have lasted maybe ten minutes stretched to two hours. "We talked about my ideas, about Irv's ideas, about the Games," remembers Godfrey. "We talked about all kinds of things. It was one of those rare and exciting moments that some people call 'flow,' or 'being in the zone;' intangible but undeniable. When we finally hung up, I instantly called my mother and said, 'I just spoke to the man I'm going to marry!'"

What Godfrey did not know was that Dardik was already married, had four children, and was almost two decades her senior. Despite those obstacles, tectonic shifts had been set in motion in both their psyches by that phone call, shifts that would change both their lives dramatically and set the stage at last for the concrete emergence of Dardik's childhood dream.

Godfrey describes herself as being fiercely driven with ambition at that time. "I was in my early 20s and in executive management. That was pretty cool, but it was never enough," she recalls. "I always had to be better, faster, badder, and anything short of being at the top wasn't good enough for me. I was doing anything that needed to be done to succeed, acting like a man, really. I wore suits and ties to blend into the role." When she had the idea of getting the pain regulators into the Olympic arena, nothing was going to stop her. "I was making dozens of phone calls, trying to find the right way to make this work," says Godfrey. "When I finally hooked up with Irv, it was one of those weird coincidences because my stepfather was on the board of trustees of Englewood Hospital."

The visceral connection she had experienced with Dardik was, however, even weirder, and irresistible—something she hadn't experienced previously, or perhaps hadn't allowed herself to experience, given her obsession with attaining corporate status at any expense. She succumbed this time. "We started talking every day on the phone," Godfrey says, "not just about my proposal or about what Irv was doing. We talked about ideas in science and philosophy. Irv loves to talk about ideas. And we were learning about each other as well."

Eventually, about a month after that first phone call, Dardik agreed to go to Minneapolis so Godfrey could introduce him to Medtronic executives and promote the plan for the pain suppressors. "I got off the plane and went to the waiting area," recalls Dardik. "She wasn't there." Godfrey admits to being a chronically late person. "So I am waiting, and then I see a woman approaching, and I say to myself, 'That's her.' It was an electric moment, but we both tried to play it cool." The two drove to Medtronic downtown headquarters in Godfrey's sporty Nissan 280ZX, a two-seater, white with a red interior—a suitable play car in San Francisco, from where she had recently moved, but hardly suitable for the approaching Minnesota winter.

"When we arrived, we talked for 10, maybe 15 minutes," says Godfrey, "and then had a working lunch with the head of the division. I thought everything was going well, and I was looking forward to the dozens of meetings that I had planned over the next four days—and, yes, expecting that Irv and I would have a chance to get to know each other better as well." Meanwhile, Dardik, who is used to feeling in control of himself, was terrified. "I went nuts over her," he explains. "I'm saying to myself, 'What am I doing here? I'm supposed to be in important business meetings, and I can't think, I can't talk.' When the lunch was over, I said to Alison, 'I have to get out of here!'"

Stunned, Godfrey drove Dardik in silence to the Pillsbury Center in downtown Minneapolis, from where he took a taxi to the airport and fled back to New Jersey. Godfrey was left with the embarrassing task of explaining the sudden disappearance of the famous Dr. Dardik. "I lied," she says. "I told them that an emergency had come up and he'd had to return home immediately. What else was I to say?!"

It wasn't that Dardik didn't want to endanger a perfect marriage. His had been rocky for a while, pretty rocky. He had fled because the power of the connection with Godfrey had been so seismic that it felt beyond his control. It was also inescapable. Inexorably, the daily calls soon resumed, with even more intensity. And in September, when Dardik went to Lake Placid, in New York's Adirondack region, to prepare for the winter Olympics the following February, Godfrey joined him.

It was a week spent in the most stunning of physical surroundings, with mountains and lakes, and forests donning their fall colors. "In the hotel, Irv played at the piano, music that he had composed himself. And then we would walk by the lake and talk about the world of medicine, which we had in common," says Godfrey. "But he would also read me poetry he had written, and then segue into science and philosophy. At some point he said to me, 'I have something to tell you that I've told no one in my life.' I couldn't imagine what it might be, so I said, 'OK, tell me.' He put his hand on his chest he said, 'Somewhere in here I have the theory of the universe, a unifying principle.'

"Even if I had tried to guess what this revelation might have been, it wouldn't have been that. So I asked him what it was, this theory of everything. His answer was equally surprising—and thrilling: 'I don't know yet, but it has to come out, and I think you can help me.' Perhaps if I had been older, I might have said, 'Come on, you're full of shit.' But I didn't. I thought he was brilliant. I was smitten."

And so, during 1980, Dardik was living multiple lives: practicing sports medicine, doing surgery, being a not-very-present husband and father, and pursuing an ever-more intense relationship with Godfrey, who still did not know about the marriage part of Dardik's life. Godfrey assembled a consortium of companies manufacturing external pain-relief equipment because she and Dardik thought that would be a more powerful way to gain acceptance by the Olympic community. "My company didn't see it that way," says Godfrey. "They fired me for 'schtupping' Dardik." Godfrey then joined a division of Johnson & Johnson as worldwide director of marketing, where she continued her macho mode of female corporate executive and moved to Philadelphia.

Whenever Dardik and Godfrey were together, which was often, Dardik would talk endlessly about unearthing the unifying principle of the universe through gaining a deeper understanding of life, which was, after all, his professional milieu. He read voraciously, in physics and biology, and yearned to spend more time in that pursuit. "I decided to quit surgery to be able to do that," says Dardik, "and so my life was divided between the Olympics, where I felt free, and reading and talking and thinking, which sustained me intellectually." He was sustained materially by license fees from the biograft, which by now were substantial.

This was the year *Cosmos*, Carl Sagan's series on the nature of the universe and of human life, aired on television. Dardik was entranced by its signature introduction: "In the last few millennia we have made the most astonishing and unexpected discoveries about the cosmos and our place within it, explorations that are exhilarating to consider," said Sagan. "They remind us that humans have evolved to wonder, that understanding is a joy, that knowledge is prerequisite to survival. I believe our future depends on how well we know this cosmos in which we float like a mote of dust in the morning sky."

Dardik saw in the message of the series something that resonated with his own quest. "He was talking about billions of stars in the galaxies," recalls Dardik, "and I saw that the image was just like the human body. The cells in the body are like stars in the galaxy, assembled together as a continuous whole. I saw the human body as a microcosm of the nature that is out there, the galaxies. I was profoundly intrigued with Sagan's perspective, and tried, unsuccessfully, to meet him. But the idea stuck in my mind."

Back down on Earth, Dardik's double life as husband and lover was proving more and more difficult to sustain, and Godfrey was growing suspicious. "When he took off for a vacation in California without me, I knew something wasn't right," recalls Godfrey, "so I challenged Irv, and that's when it all came out. I told Irv, 'You have to make a choice.'"

He did, in Godfrey's favor. Dardik extricated himself from his marriage and moved to Philadelphia to be with Godfrey. In 1982 the two set up home in the storage area above his doctor's office, on Engle Street in Englewood. Before very long, reality intruded: money. With license fees from the biograft beginning to diminish, and with the obligation of hefty alimony and child support due every month, Dardik felt obliged to go back to surgery, or at least give it a try. He also wondered whether he had made the right decision to quit. "I did three cases," says Dardik, "and I immediately knew I couldn't go back to that life, no matter what the financial pressures were. I knew I couldn't give up my search for the unifying principle."

"I did what any young woman would do," says Godfrey. "I said, 'It's all right, my darling. You follow your dreams; do whatever you have to do. I will support us.' That's what you do at 24. Here's this brilliant man, and oh, to be a part of it! I felt it hadn't happened yet for him because he didn't have the right support structure around him, and I thought, 'I can make that happen.' He had fallen in love with me because I was a full person at that time—confused, but fully functioning—and I could be there for him."

Godfrey was as committed to supporting Dardik in his quest as Dardik was in succeeding. She believed that what he was doing was of fundamental importance. And, as too often happens in this kind of situation, she began to lose herself as an individual. "I started morphing into his support structure, and I wasn't human anymore," Godfrey now admits. "I felt I had no needs, no verbalization about what *I* wanted; I just did everything for him." Gradually, this state of affairs began to be a cancer in their relationship, and, when Godfrey saw what she was doing and what Dardik was taking from her, it very nearly wrecked the marriage on more than one occasion. Eventually, the relationship became more mutually supportive.

Godfrey was struggling with who she was in other ways, too; specifically, how to *be* as a person in the business world. As the most senior woman executive at Johnson & Johnson, Godfrey had no support from other women, nor was she inclined to seek it. She was stubbornly independent. The management system was brutal and aggressive, and she was going to be brutal and aggressive, too. "It was nothing about dealing with the employees as human beings," she recalls. "It was doing anything you needed to do to get continuous and more productivity. And it didn't matter if it was eighteen hours a day, seven days a week, on the road. You burn them out, turn them over, and start again.

"I felt angry all the time, very confrontational, because it was a very angry place to be. I needed to be mean to have that aggressiveness to succeed, but it was incredibly stressful. Irv started to call me a 'corporate drone,' and that was upsetting and confusing. He was telling me, 'You are nothing but a machine,' and yet I was so proud of what I was achieving. I was so proud of who I was. But I began to see that he was right. He kept urging me to get out of there, and eventually I did. I started my own business, importing Polar heart monitors from Finland, the kind that Irv was using for his work on monitoring exercise. So I was now on my own, with my own company, no management structure. No need to be aggressive. Who was I going to be aggressive with—myself?"

Dardik and Godfrey's work lives were now more entwined than ever. "Here Alison was working with the heart rate monitors, making deals with sports-equipment companies," remembers Dardik, "and there I was, working the RITE program, using the monitors. We would both experiment on ourselves, using the heart monitors to track our pulse rates when we exercised hard, when we rested, trying to get a sense of the rhythms of our bodies. I was beginning to read a lot of books in biology for the first time, books like Lewis Thomas's *The Medusa and the Snail*, talking with Alison about it. I realized that a lot of writing about nature is about the idea of *controlling* nature. It's in the Bible, dominion over nature, that kind of thing. What is this thing about control? It's everywhere, even in medicine. We talked about all this, saying that we have to trust that there is something in nature that we are missing, that we refuse to understand or even want to think is there. I knew that nature had something to tell us, something like the *elan vital*, or life force, but it would be a concrete phenomenon, not mystical. I just had to find out what it was, that's all."

Dardik and Godfrey thought of themselves as the king and queen of heart monitors, and they got their friends to exercise with them, too. One such friend was Jack Kelly Jr., who, like his father, was an Olympic rower. The senior Kelly started life as a bricklayer from Ireland and went on to own his own construction company, make a lot of money, and become a big presence in the politics of his American hometown of Philadelphia. Despite being an Olympic gold medalist—in the single sculls in 1920, and the double sculls in 1920 and 1924—he was barred from entering England's upper-class Henley Regatta because of his working-class background.

Jack Kelly Jr. avenged that bit of English snobbery when he beat the Brits at Henley in 1947 and 1949, taking the gold both times. He also won a bronze medal in the Olympic single sculls in 1952, a world title in 1949, and numerous other titles. East River Drive, a scenic route that is boarded by the Schuylkill River on one side and Philadelphia's beautiful Fairmont Park on the other, was renamed Kelly Drive in his honor, complete with a bronze statue of his father rowing.

Kelly attended the University of Pennsylvania, like Dardik, but the two men didn't meet there. That happened years later, when Dardik was pursuing sports medicine with the Olympics and Kelly was vice chairman of the U.S. Olympic Committee. The two men became close friends and spent time together when they could. They went to the winter Games in Sarajevo in 1984, and took time off from their duties to relax together. "Alison and I got Jack to wear a heart monitor when we were running in the mountains around Sarajevo," remembers Dardik. "Jack had a resting heart rate of 40 beats a minute. That's incredibly low, supposedly a sign of endurance capacity and super fitness. He would go out for long runs, not sprinting, and we would sprint past him, then wait for him to catch up; then we'd sprint again, doing our intervals, monitoring our heart rate. We had a lot of fun together."

But Dardik had a nagging unease about his friend. "I had an uncomfortable feeling about distance running," he recalls. "I loved sprinting and couldn't imagine anyone wanting to run distances. But it was more than that. I had seen a lot of world-class distance runners have a lot of health problems, infections, that kind of thing, and I felt it could be dangerous, even fatal. In Sarajevo I was constantly ribbing Jack, saying things like 'You're going to kill yourself, running like that.' Jack just laughed and said that he liked running distances and that it was good for his body."

As we know, Dardik's unease was well founded.

The evolution in Dardik's thinking—about health and exercise and the fundamental nature of nature—that had been triggered by Jack Kelly's death flowed with intoxicating ease and at great speed. "It was just like that moment in the operating room at Montefiore when I realized that the human umbilical cord was the answer to the bypass problem," recalls Dardik. "It was all there in a flash: that cyclic exercise was natural and would enhance health, and that waves were the fundamental stuff of nature and the universe, at all levels." So excited was he with this early-hours insight that he immediately phoned his sister Sylvia at four in the morning and said, "Hey, Syl, it's waves. *It's all waves.*"

Dardik's ideas about waves would undergo refinement over the years, of course, but essentially the die was cast for his intellectual journey. For much of the subsequent decade, beginning in 1985, the ex-vascular surgeon seemed ever more ready to challenge conventional medicine and conventional science. It was, however, to be a wrenching roller-coaster ride.

The decade began with a setback—at least it seemed that way initially. When Jack Kelly was made president of the U.S. Olympic Committee, he shared Dardik's concern about the rising use of drugs in sports and an emerging practice of achieving improved performance by other dubious means. Athletes found that if, shortly before their event, they had a transfusion of blood, they could boost their physical execution. The reason was simple. Known as blood doping, the act of adding blood to an individual's bloodstream delivered more oxygen to the muscles during exertion. Greater oxygen availability translates to more intense and efficient muscular activity. On the sports field, that leads to faster and more sustained running, for example.

Blood doping had been outed in the 1984 Olympic Games in Los Angeles. Many U.S. athletes were involved, and there seemed a possibility that some Olympic officials knew what was going on and had turned a blind eye to this and other irregularities. As chairman of the U.S. Olympic Sports Medicine Committee, Dardik was charged with producing a report on the blood-doping scandal and drug use in general, aided by three physicians associated with the Olympics.

Jack Kelly was never able to act on the findings of the report, which was finished just days before his death. Instead, the report—a preliminary investigation, really—went to Kelly's successor, Robert Helmick, who had been a U.S. Olympic Committee vice president. "We couldn't prove any specific irregularities," recalls Dardik, "but we did find enough evidence that further investigation was essential, and we said so plainly in the report." The end of March came and went, weeks after Dardik's report had reached Helmick's desk. No response. Shortly after that, Dardik received in the mail a plaque that thanked him for his services in bland bureaucratese.

Had Jack Kelly not died, events might well have unfolded differently for Dardik, and for drug use in sports in general. But as it was, Dardik was ready to exit the Olympic scene, frustrated by not having made a greater impact on athletic performance, especially in the context of rampant drug use; aghast at the "winning is the only thing" attitude that was increasingly dominating Olympic participation; and saddened by the abrupt extirpation of a promising professional partnership with his friend Jack Kelly.

With no Olympics-related activities to occupy him, and with his surgical practice now five years behind him, Dardik withdrew once more into his "cave" and furiously focused on nurturing his nascent exercise protocol and his theory of everything. "Getting out of the Olympics was a blessing in disguise," says Dardik. "It freed up my time so I could concentrate on my real passion, the fundamental importance of waves in nature. That came out of seeing the HeartWave that was prompted by thinking about Jack Kelly's death. And it was strengthened by my recollection of a guy called Cooper, on whom I had operated years earlier."

In the fall of 1974, Cooper, a man in his 50s, was rushed to the emergency room at Montefiore Hospital, showing symptoms of a perforated bowel. Dardik was on duty, and he opened the patient up. "Unbeknownst to anyone, the guy had colon cancer, with metastases all over the peritoneum, the liver," recalls Dardik. "He looked like a real goner. I removed the tumor, did a colostomy, closed him up, and expected that he would die pretty soon." Cooper had widespread infection, and over the next couple of weeks his temperature spiked repeatedly, up and down, up and down. To everyone's amazement, Cooper didn't die, and eventually he was discharged. Dardik was no longer his physician because he was transferred to a cancer specialist.

Almost a decade later, during Yom Kippur of 1983, Dardik was at temple in Englewood. A man came over to him and said, "Dr. Dardik, do you remember me?" Dardik looked at him for a moment and said, "Cooper? Mr. Cooper, the man with the colon cancer?" The man nodded. "I was in shock," says Dardik. "The guy should have been dead. He had had metastases all over. I said to him, 'What happened to you?' To which he replied, 'Nobody knows. There was a spontaneous cure.' It was a miracle, but it immediately made me think of Colley's toxin."

William B. Colley was a surgeon in New York at the turn of the century. He became intrigued by the observation that some cancer patients with bacterial infections had spontaneous cures. He thought that the toxins produced by the infecting bacteria somehow killed the cancer cells, and he was pursuing the use of toxins in cancer therapy. Then chemotherapy came along and this line of inquiry was dropped. "I saw it differently," says Dardik. "What I saw in Cooper was that the repeated high temperatures he had were making waves in his body, and that somehow normalized the cancer cells."

Researchers at the National Institutes of Health had repeatedly tried hyperthermia, or raised body temperature, as a cancer therapy: 108°F for as long as the patient could take it. "That's like running a marathon," says Dardik, "and it caused DNA damage, so they stopped. It's the wave-like repetitions of high and low temperature that cause health, just like bursts of exertion and recovery cause health." Dardik became obsessed with making waves in relation to health.

He read voraciously, principally in physics, especially quantum physics, and was fascinated with cybernetics, the study of feedback systems, which, he thought, captured the essence of what he was seeing in the repeated, linked cycles of exercise and recovery. So intrigued was Dardik with the parallel ideas in cybernetics and cyclic exercise that he coined (and copyrighted) the term Cardiocybernetics for his exercise protocol, a description of which he had drafted within a year of Jack Kelly's death. The paper, "Cardiocybernetics: Relaxation through Exercise," was published in *Advances: Journal of the Institute for the Advancement of Health* in the summer of 1986. The Cardiocybernetics program involved bursts of intense exertion, interspersed with recovery, as many as 20 times over a period of an hour.

The rationale for the need for Cardiocybernetics, Dardik explains in the paper, is that the dynamics of modern life deprive most people of the opportunity to relax; the physiologic changes associated with relaxation—lower blood pressure, slower pulse, reduced oxygen consumption—have withered through lack of use, "much as a muscle atrophies with disuse." When humans lived in harmony with nature, we routinely found ourselves faced with the urgent need to respond to dangerous situations, such as in the hunt or when being hunted. The fight-or-flight response kicked in, raising blood pressure, pulse rate, and oxygen consumption. These are the physiological correlates of stress. When the challenge had passed, relaxation was a natural state to adopt, with the concomitant reversal of cranked-up physiology. This recovery process deactivates the fight-or-flight response, completing a natural cycle that all humans experienced day by day, year by year, and generation by generation. Until today.

People in modern society no longer face the prospects of hunting and being hunted, then relaxing; we no longer experience the wave of stress and recovery physiology, as we once did. Making waves through cyclic exercise is a way of returning human physiology to a more natural, healthy state, Dardik hypothesizes, ending the article as follows: "Cardiocybernetics is . . . a program that will enable investigators and practitioners to explore its use in enhancing human performance and preventing disease."

Cardiocybernetics is an evolution of the Rhythmic Interval Training Exercise that Dardik developed half a decade earlier for Olympic athletes. "RITE was essentially interval training, where I had athletes run hard for a short time and then walk to recover toward a normal heart rate. With Cardiocybernetics, you stop completely after the burst of exertion, and even sit down. The heart rate recovery is much faster and deeper that way," he explains.

The way that new element in the protocol came about was simple, and circumstantial. Alison Godfrey was pregnant with her son, Trevor, during 1985, the year she and Dardik married. She was still doing interval training with Dardik throughout her pregnancy and found it much easier on her body if she sat down after each bout of exercise. "I was wearing a heart rate monitor, of course," remembers Godfrey, "and I saw a huge improvement in recovery. I said, 'Wow, look at this, Irv. This is great. You should try it.' He did, and he saw the results. That's how stopping completely by sitting became part of the protocol."

At first, Dardik and Godfrey used the cyclic exercise protocol on just themselves and immediate family. But in January 1987, a healthy couple from Englewood, who had heard about the protocol from Dardik's brother, Herbert, said they wanted to try it. They did it for a few months, loved the rhythmicity of the experience, and claimed that they felt much healthier and more at ease with themselves mentally. The first inclination that the protocol could have the positive impact on disease that Dardik had hypothesized came from Dardik himself and his own medical history with ankylosing spondylitis. "I was able to keep it in check with drugs, steroids, but it was still there, and it would flare up every so often," Dardik says. "I was always in pain. I had sciatica in my legs, arthritis in my neck, and I always had a low-grade fever. There's nothing you can do about it but take drugs.

"Then, during 1987, I began to realize that the symptoms were getting less and less, until they finally disappeared. Completely gone. That told me that doing cyclic exercise really worked in reversing disease. But if I stop exercising for any length of time, I can feel the symptoms creeping back. You have to keep doing the cycles to maintain health."

The old phrase "Physician heal thyself" appeared to apply to Dardik and his disease. Except that he wasn't doing it as a physician. He was doing it by making waves through cyclic exercise.

Remarkably, Dardik found that another affliction he suffered was also ebbing: a paralyzing fear of flying. Just the prospect of getting on a plane drove him into panic. Within a year of starting regular cyclic exercise, the phobia was gone. "By doing exercise and recovery," explains Dardik, "I had trained my physiology not to automatically react to fear by flooding my system with stress hormones. By training recovery, my physiology was able to handle the stress of fear more effectively. A vicious cycle had been broken."

Very soon, word got around about what Dardik and Godfrey were doing and, as Godfrey puts it, "everybody with a medical problem started to come out of the woodwork, asking to do the program. We were getting terrific results, with people with all kinds of conditions, from multiple sclerosis to anorexia." Financial pressures were building up in the Dardik household at the time, though. Dardik had no income, and Godfrey's business of importing Polar Heart Monitors from Finland collapsed when Polar decided to do the importing themselves. And there was the alimony to Dardik's ex-wife. "We needed to earn money, so we decided to make it into a business, charging people what they could afford," Godfrey says.

Dardik and Godfrey were living on a 45-acre farm near Hackettstown, in western New Jersey, that they had bought in the summer of 1985. It was a rambling 200-year-old house on a hill and had been added to over the years, with exposed beams and tiny rooms. There was a huge barn of the same vintage, along with stables for a dozen horses. A small cottage a hundred yards from the main house added more space. The house was surrounded by copses and fields—an idyllic setting, especially for kids. But it was impractical from a business point of view, as most of the clients were from Manhattan. Getting to Great Meadows was not easy.

Dardik and Godfrey therefore decided in 1987 to rent an apartment in New York, on 43rd Street, a few blocks up from the United Nations building. "It was a tiny one-bedroom place, with the elevator shaft opening up into the bedroom," explains Godfrey. "Trevor was 18 months old, and I was heavily pregnant with Whitney. What a scene it was! People would come to the apartment and start doing cycles at six thirty in the morning, in the tiny living room. Trevor would be playing in the kitchen sink."

Dardik and Godfrey had half a dozen clients, including "Michael," who suffered from Tourette's syndrome, a neurological disorder that causes repeated involuntary movements or vocalizations. In Michael's case, it manifested as vocal whoops and curses every six or seven seconds. "I remember one evening when he was with us," says Godfrey. "He was in the living room doing cycles, and I was in the kitchen with Trevor. The whoops would come drifting into the kitchen. All of a sudden, Trevor started to whoop back, mimicking Michael. I was mortified, but Michael said, 'It's OK. Children see me as being playful, just another child they can relate to.'"

Within a few months, Michael's whooping was down to every 20 seconds or so, and he was getting better with every passing week. "One day he came to me and said, 'You know, Irv. I'm going to have to stop the program,'" recalls Dardik. "I need my disease back." Michael was a television personality, very much in demand, and when he lost his symptoms, he was just like anybody else. He preferred to be special with his symptoms, rather than just another guy without them.

Whitney was born in January 1988, and living in the apartment was no longer practical as a home and workplace. Godfrey, now the mother of two young kids, insisted on a change. A move back to the farm was planned, but with a difference this time: clients would come and live there, often for months at a time. "The reason we decided to do it that way," explains Dardik, "was that the teenagers we were working with during the day would go back to their families in the evening and fall right back into their old patterns. These were mostly kids with psychiatric issues. We concentrated on these kinds of behavioral problems in the beginning because I see them very much as problems of wave patterns, which cyclic exercise is designed to renormalize. It would be lights out early, up early in the morning, get them to do cycles on our schedule, get them to eat on our schedule, make them take naps. So we needed to control their entire environment.

"One was a young man who was psychotic and violent, and who on one occasion tried to run Alison down in a car. He felt ugly and deformed, even though he was good-looking, and constantly hid his face behind his hands to escape being seen. Another was a young woman with anorexia, who felt she was ugly and fat. But after a few months on the program, both of them recovered enough to go back home and live a normal life.

"I remember when George came back to visit us one day. He walked up to the porch, smiling, wearing sunglasses. He said, 'Hello,' shook our hands, and we hugged. He told us he was holding down a regular job, something that had been impossible for him before. Cynthia graduated from high school and went to college. Her mother called us and said, 'I can't believe this. My daughter is out on the beach in a bikini! Thank you.' It was a beautiful thing. We had so many stories like that."

One such was "Eric," newly married, with severe schizophrenia. "He had bad hallucinations and would often be in a fetal position, screaming and crying," recalls Godfrey. "He was completely dysfunctional: couldn't work; couldn't hold his marriage together. He stayed with us for a couple of months and did extremely well, and then moved back home." Some time later, Eric called Dardik, agitated. "I am extremely upset," he told Dardik. "I'm hallucinating again. I can see a six-foot snake in my bedroom!"

"Are you sure you're hallucinating?" Dardik asked, puzzled. Eric had done so well on the program, and his heart rate numbers were still extremely good. "Could there be a snake in your room?"

"Of course not; it has to be an hallucination."

The two men went back and forth like this for a while, until Dardik finally persuaded Eric to call the police to investigate. Sure enough, there *was* a snake in Eric's room. It had escaped from somewhere, no one knew where, and ended up in Eric's apartment.

Despite the successes, the financial straits the couple had been laboring under for a while became yet more urgent. Godfrey felt burdened, taking care of two children and a husband, cooking three meals a day for the clients and her family, looking after the horses, delivering the program to half a dozen clients, and from time to time having to barricade her children and the dog in an upstairs room when a client became upset and violent. Dardik, meanwhile, was becoming less and less involved with the day-to-day business of working with clients and spending more and more time in the barn, surrounded by mountains of books and papers, working on his theory. "I was consumed with my work, and I was leaving too much to Alison," admits Dardik. "But I felt driven because everything was falling into place. When I decided to go to Pittsburgh to write up the theory with my sister, it was very hard to explain to Alison that I was taking off, leaving her to take care of things. All her friends were saying, 'What is he doing?' It was as if I were divorcing myself from the family."

During his two months in Pittsburgh, Dardik met frequently with his friend Greg Andorfer, the producer of Carl Sagan's *Cosmos* television series, to mull over ideas. During their conversations, Dardik more than once related the apocryphal story of the man seen at night looking for his car keys in the light under a streetlamp. Asked where he thought he had lost his keys, the man says, "Over there," pointing to a dark, unlit part of the street. "Why, then, are you looking here and not over there?" he is asked. "Because this is where the light is!" the man replies, exasperated. Dardik would say to Andorfer, "That's what modern science has been doing—taking the easy but misguided way out. The reason I've been able to come up with something completely new is that I've been looking in the difficult, dark, but fruitful places."

By the end of his two-month sojourn in Pittsburgh, Dardik's paper was finished, a 20,000-word scientific epic. He had a thousand copies printed and bound and sent copies to a handful of physicists around the country. The result was pretty much a deafening silence. "I wasn't really surprised or discouraged," Dardik recalls. "What I am saying challenges everything in science, so I expect scientists to disagree with me, initially anyway." Marvin Goldberger, the director of the Institute for Advanced Study, in Princeton, was, however, sufficiently intrigued to agree to talk to Dardik.

"I met him in his office, which was the same one that Einstein used," Dardik recalls. "He said, 'You can have five minutes.' So I started talking rapidly, and I asked if I could draw on his blackboard, Einstein's blackboard! It was covered with all kinds of equations. He said, 'No, that's for a seminar I'm giving this afternoon.' Five minutes go by. Ten minutes, and I'm still there. He then said, 'OK, go ahead and use the board.' He asked a lot of questions, and I answered them. We went on like this for two hours, and he seemed interested, but I never heard from him again. That often happens. People like what I tell them, but then I go away and they go right back to their original thinking."

The same thing happened with Ilya Prigogine, who won the 1977 Nobel Prize in chemistry for his work on nonequilibrium thermodynamics, showing that molecules communicate better when they fluctuate. He also said that cells and molecules are healthiest when they are oscillating, or making waves, an obvious parallel with Dardik's theory. "We talked for an hour on the phone," says Dardik. "And then I heard nothing more from him."

Dardik had a more positive response from David Bohm, one of the most distinguished physicists of his generation. A phrase in Bohm's book *Wholeness and the Implicate Order* caught Dardik's attention because it sounded so similar to what he was saying: "We must learn to view everything as part of undivided wholeness in flowing movement." Dardik went to England to meet Bohm and stayed at his house for a couple of days. "We talked for hours," recalls Dardik, "and he said, 'You're right. I think you're right.' He had some reservations, of course, but he was very supportive. After I came back to the States, we talked on the phone a few times, all very positive. And then he died, in 1992." Dardik realized he had to continue to be patient.

When Dardik returned to the farm from his paper-writing stint in Pittsburgh, he was in high spirits, his theory finally formulated and in print. Godfrey, meanwhile, was most definitely not. The financial crisis was deepening. The bank was threatening to repossess the house. And the stress of maintaining the house and working with clients was becoming ever more acute. Breaking point came one day in March 1990, when Godfrey came downstairs to find her daughter in the room with one of the clients, a young woman with obsessive-compulsive disorder. "This young woman would do sit-ups for hours on end," explains Godfrey. "And there was Whitney, lying on the floor next to her, copying her, doing sit-ups, this little girl! I said, 'I am so done with this.' I packed my bags, got the children together, and left to stay with my sister, Neale."

Everything at the farm fell apart. The clients soon left, as Dardik was unable to maintain their programs or the house. He did, however, continue to work with clients away from home. Before very long, Dardik left the farm, too, so that Godfrey could return with their two children. He stayed at Neale's house for a short time and then variously lived out of his car, crashed on friends' couches, or slept on a park bench in Florida, where he was working with a young client. Divorce was in the offing. It was a dark, dark time.

PART TWO

For Better or Worse, Everything Is Waves

7

Emotional Waves

"We shall not cease from explorations
And the end of all our exploring
Will be to arrive where we started
And know the place for the first time."

—T. S. ELIOT

THE CALL CAME OUT OF THE BLUE, in the late summer of 1990, while Irving Dardik was still living a nomadic life. "Hello, Dr. Dardik, I'm Tony Schwartz," the voice on the phone said. "I'd like to interview you about your work on cyclic exercise and chronic diseases, for an article in *New York* magazine. "I was very cautious," recalls Dardik. "I had had interviews before, when I was with the Olympics, and I know they can be destructive. Here I was, quietly working away on developing my theory further, and here was this guy I didn't know who wanted to write about me. I suspected he thought I was a fraud, and he just wanted to expose me to the world. But he was very persuasive, and I eventually agreed. I said to him, 'OK, I'll meet you, but on one condition—that you listen to *everything* I have to say, not just the exercise and health part.' He agreed."

Schwartz was a writer of some stature, having recently coauthored the best-selling book *Trump: The Art of the Deal* with Donald Trump. Schwartz had heard about Dardik's work from his friend Jim Loehr. A sports psychologist, Loehr had met Dardik while heading sports medicine for the United States Tennis Association when Dardik was still involved with the Olympics. Loehr was very impressed with the Dardik's Rhythmic Interval Training Exercise program for Olympic athletes, and he incorporated Dardik's ideas into his own program for training tennis stars. The two men had kept in touch over the years, so Loehr knew of Dardik's work with people with chronic diseases, and he told Schwartz he thought it might make interesting material for an article. Schwartz was close to finishing a book he was working on at the time, *What Really Matters*, an exploration of mind-body relationships in self-discovery and self-improvement. "When I heard about Dardik's work, I was very intrigued," Schwartz says, "so much so that I was prepared to put the book aside for a while and find out whether Dardik was on to something."

The first time the two men met was at Schwartz's home. "We went to my office, and Dardik immediately launched into a dazzling exposition," remembers Schwartz. "He didn't sit down. Instead he paced back and forth for two and a half hours, gesticulating and talking about physics, astronomy, quantum physics, about the similarities between top athletes and people with chronic diseases, and about his exercise program and how it could reverse chronic diseases. I'm not a scientist, so it was hard for me to evaluate what he was saying about the science. But what he was saying about exercise and recovery made a lot of sense to me. He was obviously a man with big ideas, but whether they were stupid big ideas or brilliant big ideas, I couldn't be sure. But they were fascinating enough for me to want to find out. I spent the next four months with Dardik, pretty much every day, as he worked with some of his clients, and I interviewed other clients separately."

A couple of months into the interview process with Schwartz, Dardik and Godfrey came to a reconciliation over their marriage problems, and Dardik moved back to the farm on Thanksgiving. There was one important condition on the reunion: Godfrey insisted that never again would the farm be home to half a dozen clients.

The article "Making Waves: Can Dr. Irv Dardik's Radical Exercise Therapy Really Work Miracles?" appeared in the March 18, 1991, issue of *New York* magazine and was featured on the front cover, with a picture of Dardik, clad in a white lab coat, dumbbells in hand, jumping on a trampoline and smiling broadly. The article not only presented the science in a detailed and considered manner, but it also described the exercise protocol and its results in about as laudatory a tone as Dardik could have wished. "I soon confirmed that Dardik has had some extraordinary results treating patients with illnesses ranging from anorexia to multiple sclerosis," wrote Schwartz in the introduction to the article. "He's yet to conduct any clinically controlled trials, but his anecdotal results are impressive. While physicians are often at a loss to explain why a given treatment works and scientists have yet to find a cure for any chronic disease, Dardik's explanation for his effectiveness was, at a minimum, plausible, provocative, and intuitively compelling."

The core of Dardik's approach, as Schwartz described it, is that health depends on a balanced relationship between stress and recovery, each of which is unhealthy by itself. "Dardik speculates that when a person makes large waves of energy expenditure and recovery," wrote Schwartz, "the body's immune chemistry and repair processes are activated. In turn, they're prompted to make their own healthy waves and to do their work more efficiently." Schwartz noted that Dardik drew support from the work of Nobelist Ilya Prigogine, who argued that living systems, at all levels, are healthier when they are oscillating, or making waves.

Schwartz further explained that people with chronic diseases typically cannot raise their heart rate as high as healthy people can when they exercise intensively, and that one of the aims of Dardik's exercise protocol is to elevate maximum heart rate to restore healthy waves throughout the body's physiological systems. Dardik, Schwartz explained, uses heart rate as a window to the overall physiology. He then told the stories of a handful of Dardik's clients, beginning with Marianne Rouis, a 48-year-old housewife and part-time interior decorator. She had been suffering from severe enteritis—inflammation of the intestine—which meant that she often couldn't stay away from the bathroom for more than a few minutes at a time. "The problem goes back to 1971, when I had an emergency C-section for my third child," Rouis told me. "Apparently I was acutely sensitive to a drug they gave me during labor, Pitocin, to stop bleeding. It felt like my body was going a hundred miles an hour, not just my pulse but my whole body. It felt as if someone had injected me with amphetamines. That happened every time I was stressed, and eventually it affected my GI tract. They told me it was my nerves, but I knew it wasn't. Then a friend who had had good results with Dr. Dardik suggested I call him, and I did, even though I've never done anything like that before."

Rouis went to Dardik's farm and stayed for a couple of weeks. "He spent a lot of time going over my history," says Rouis, "and eventually concluded that Pitocin had triggered the enteritis. He said, 'I think I can help you.' I started doing cycles with him, 20 in the early morning, 20 in the late morning, 20 in the afternoon, and 20 in the evening. It was pretty experimental back then, and he was still learning." Rouis fully committed herself to the program, and she even stopped her work as an insurance broker for two years to do it properly. "I made use of that time at home, to become certified in interior design, which is what I had always wanted to do," she says. "Very soon into the program, my heart rate began to normalize from its previously very high level, and I could get it up with exercise, and then bring it down with recovery. And the symptoms of enteritis started to go away, and then disappear completely, after 10 weeks. It was like Dr. Dardik had given me my life back."

Schwartz quotes Rouis as saying, "Something just clicked in my body, and my whole life changed. You get very connected to exactly what is going on inside. . . . When you are in tune and you are getting a really good swing up and down, it's almost a feeling of euphoria."

The second case Schwartz described was of George Hornbeck, who had recently retired from the plumbing and heating business. Hornbeck was a good friend of Marianne Rouis, and it was he who had recommended that she go to Dardik. Back in 1979, Hornbeck had run the New York marathon, and when he got home he collapsed at around 5:00 P.M. He was rushed to the local hospital in his hometown of Monticello, where doctors were unable to determine the problem until about 11:00 that night. A clot was cutting off the blood supply to his left leg.

The doctors worked throughout the night, to try to break up the clot. "They came to me around noon and told me that they were going to have to amputate the leg, at the thigh," Hornbeck's wife, Shirley told me. "I was appalled. I was crying. But I talked to a lot of the doctors and asked, 'What would you do if it were *your* husband?' I was very involved with the hospital, so I knew these people. We decided that he needed to get to the best doctors, so George was transferred to Englewood Hospital."

Dardik was on duty and, knowing that Hornbeck was a runner, immediately opted to do a Syme amputation, which involves amputation of the foot, leaving the heel in place as a foundation to walk on with a prosthesis. "I was so happy Dr. Dardik was able to do that," recalls Shirley. It was, however, the beginning of a miserable few months for Hornbeck, who remained hospitalized for several months and underwent more than 20 further surgeries. Even when he was discharged, the problems were not at an end, and in fact got worse. He began to suffer phantom limb pain, a bizarre and largely mysterious phenomenon in which an amputee suffers pain, often severe, in a limb that is no longer there. Hornbeck was taking heavy-duty painkillers and soon became addicted to them. To combat depression he was put on huge doses of the antidepressant Elavil, 450 milligrams a day. Life was awful and continued to be so for several years.

Then, in 1987, his physician, Herbert Dardik, suggest that Hornbeck seek relief from his problems by seeing his brother, Irving Dardik. At first, Hornbeck could reach a maximum heart rate of only 108. Within a few months he could hit 145, his depression started to lift, his addiction abated, and his pain eased. Shirley, a teacher, took a year off from work to do the program with her husband. Toward the end of the year, they went to the Caribbean to spend some time with friends. "Shirley called me," recalls Dardik. "She was so excited. She said, 'Not only has the depression gone, but the pain has gone as well.' They came back to the States, continued to do cycles, and George was able to go back to his work for the first time in years."

During the course of the year, as his depression eased, Hornbeck began to cut back on his meds. "He was still attending a depression clinic," says Dardik. "They looked at his blood and saw that the level of Elavil was far lower than they had prescribed. They said, 'You can't be feeling better because the level of Elavil is so low.' George asked me to talk to them, so I called up, and they said, 'He's in trouble. He's going to commit suicide if he doesn't take his meds.' I explained the program, but they wouldn't listen. 'You can't reverse depression with exercise,' they said. I said, 'Well, he's getting better, isn't he?' They still wouldn't listen."

Schwartz described Nancy Kaehler (now Kaplan) as the most dramatic story of all those he heard. Kaplan had been diagnosed with multiple sclerosis early in 1989 and, as Schwartz wrote, "had been on a relentless downhill course until she met Dardik." Kaplan lost feeling on one side of her body; was catheterized because of failed bladder function; suffered from migraines, general weakness, exhaustion, lack of sleep, and skin irritation—all of which added up, she says, to "constant discomfort." But she was not in a wheelchair. "I was never willing to accept inability, to be immobile," she now says. "I would lean against walls when I moved, and I have all kinds of bruises to prove it. I had accepted the fact that I had MS, and I had accepted the fact that there isn't a cure for it. But I wasn't going to let it stop me."

When, in August 1989, her mother said that her oral surgeon, Elliot Dardik, had told her that his brother Irving could help reverse those symptoms, Kaplan was skeptical. She did, however, agree to talk to Dardik on the phone, more to please her mother than expecting any prospects of help. "He told me about his theory and used analogies about how my physiology was like a symphony orchestra out of sync, that kind of thing," she recalls. "I said something to the effect of, 'Hey, buddy, how much is this going to cost? I'm not going to mortgage my house.' He wouldn't tell me, so I blew him off."

Kaplan's mother was relentless and said, "I have never asked you to do anything for me before. Do this one thing, please!" Kaplan gave in to her mother's pleas and agreed to meet Dardik, on the condition that her cousins Jack and Bobby go with her.

Kaplan and her entourage met Dardik later that week, and he spent two and a half hours talking about his ideas and the exercise program. "By the end of that time," Kaplan recalls, "I said, 'Look, I have an appointment to go to. When do we start?' That was it. It made sense to me, so I just wanted to get on with it."

Kaplan was an employment agent, dealing with financial institutions, but by October she realized she had to give that up so she could devote the necessary time to the program. One day she was cooking with a friend in her parents' kitchen, making red sauce in an eight-quart casserole. "At one point I lifted the heavy casserole off the stove and carried it to a nearby table," recalls Kaplan. My friend looked at me and said, 'Oh, my God! Do you realize what you just did?!' She was aghast. 'You haven't been able to do anything like that for more than a year.' She was right, but I didn't really notice. It was often like this, other people pointing out that I was getting my strength back."

When Kaplan had started on the program, she was so weak that holding something as light as a glass of juice was a major two-handed effort. By December, she was growing stronger by the day. She could walk without needing support from walls, she no longer had to be catheterized, her migraines were gone, and she could sleep. Then, on January 2, her beloved grandfather died. "I was devastated," Kaplan recalls. "But I was the one making the funeral arrangements, ordering the food, ordering the flowers. I thought I was doing fine. But the day of the funeral I crashed physically. I was so weak, I could barely walk to the bathroom in my house."

Dardik recommended that Kaplan spend some time in a warmer climate, away from reminders of her grief. So she went to Los Angeles to stay with her sister for five weeks, rapidly regaining her strength and her emotional energy. And to prove it, there is a photograph of Kaplan in the *New York* article, jumping on a trampoline, smiling, at her sister's house.

Kaplan returned home at the beginning of March and made a decision: "I said to myself, it is time to fit the program around my life, rather than have my life fit around the program." She made another decision, too. In the time before Kaplan met Dardik, she had given up dating "because it was too complicated." But when she got back from Los Angeles, she started dating again. "I dated a couple of guys, and then I met Josh," she says. "We were married in June." This was four months after Schwartz's article had appeared. After a honeymoon in the Greek Islands, Kaplan went to Costa Rica for two weeks with Dardik and a small group of his clients. "We would get up at 4:30 in the morning and do cycles," she says. "And we would go to bed really early, around 7:30. I felt great. I kept up that schedule when I got home, and very soon I felt I had the strength of 10 people! I said to myself, 'OK, it's time I had kids.'"

Sophie was born in 1996, and when she was six months old, Kaplan became pregnant again. "It was the dumbest thing I ever did, because Irv and Alison had told me that one kid was enough," she says. "I crashed, to the floor. I gradually recovered over the months, but not to the point that I was at before I had Sophie. But, you know, I will get there soon."

Schwartz quotes Robert Sergott, a neuro-ophthalmologist who treated Kaplan, as saying, "You always have to question how much of these results might be a placebo effect. . . . Until you see controlled studies, it's terribly difficult to know how much of the healing effect to attribute to any specific therapy." Schwartz also reported Dardik's own experience with the cycles reversing his symptoms of ankylosing spondylitis.

Nicholas Hall, a psychoneuroimmunologist from the University of South Florida who has published more than 100 scientific papers in fields ranging from psychology to immunology, was distinctly positive about Dardik's work. "I've become increasingly convinced that Dardik is really on to something," he is quoted as saying. "The concepts embodied in his theory are so simple that it is difficult to imagine that they might contain the answers to many complex biochemical questions, but they make enormous intuitive sense. They still must be tested, but I think his ideas will eventually replace traditional thinking in biomedical circles."

Hall has had extensive exposure to Dardik's thinking and his work with clients. In 1987, Dardik read some of Hall's work, liked what he saw, and immediately called him on the phone; this was while Hall was at George Washington University in Washington, D.C. "We had a long conversation, and we really clicked," recalls Hall. "Irv got on a plane that same day and came down to see me. That's very typical of Irv." Thus began a collaboration that continued for half a dozen years. "Irv is very passionate and committed to his theory," Hall continues. "He gives every penny he earns to people like me and others, so they can do research to extend and refine the theory." Describing why he resonates with the thrust of the theory, Hall says, "It has an elegant simplicity. Irv's genius is that he is able to bring lines of thinking from many different disciplines and make sense of them as a whole, giving an entirely different way of looking at, in this case, health and disease."

Hall is now director of wellness at Saddlebrook Resort, a corporate retreat in Tampa, Florida. "I continue to use Irv's principle in my own work," he says, "and I know other people have, too, such as Jim Loehr, in his work enhancing performance with corporate executives. There is something very powerful in making waves."

Schwartz closed his article as follows: "Dardik's unified theory isn't likely to win wide acceptance by the scientific Establishment anytime soon. What's harder to dismiss is his clinical success in treating patients with chronic diseases. If he can demonstrate that his program is as effective for patients in controlled studies as it has been anecdotally, even skeptics will feel compelled to take a closer look at his broader ideas. Dardik's future rests with his patients—and his patience." He would certainly need a lot of the latter, and now there was no shortage of the former.

Dardik was delighted with Schwartz's article, but he had no idea what it would provoke among readers. "It was just like when *Parade* ran the story about the biograft, only much more so," he recalls. "We were *swamped* with people wanting to do the program, thousands of them, but we could only take a few." One of these was a man named Morty, whom I met on a hazy July day, in his 41st-floor apartment on Manhattan's West Side. Pretty much the first thing he said to me was, "I am 76, and I don't look it, and I don't feel it." It was true. He looks at least a decade younger than his years.

Born in Brooklyn in 1928, Morty lost his mother when he was just five years old. His father, who owned a roofing company, soon remarried, and Morty was thrust into a very contentious relationship with his stepmother—there was lots of animosity. "That experience taught me to be independent at a young age," he now says. His brother, a couple of years his senior, went to Syracuse University in New York, where he studied structural engineering. Morty followed in his footsteps and was a member of the university's boxing team. The two brothers went into the engineering industry, building bridges and tunnels, after leaving Syracuse.

In 1956 Morty's father had a heart attack, the first of what would be a series of such events over the next 20 years. Morty felt compelled to take over his father's business for what he thought would be a temporary period. He discovered that the company was in a fragile state financially, and he had a tremendous struggle to prevent it from going under. He also discovered that the commercial roofing business was, as he puts it, "the asshole of the construction industry." The quality of work was universally abysmal, jobs were assigned to the lowest bidder, and making a profit was the exception. He decided that the path to success would be through delivering a top-quality product for which contractors would be willing to pay good money. It worked, and Morty, joined by his brother, built what he describes as "the biggest and the best roofing company in the business." Pointing out the window, he said, "Most of the big office buildings out there, from the '50s, '60s, '70s, we put the roofs on them."

Despite the company's success, Morty was restless, leaving more and more of the business for his brother to run while he dived into a lot of other ventures, including in the financial world and, most recently, the oil-servicing industry—all very successful. He also began to invest in and subsequently produce plays, an avenue into which his actress wife led him. Being a successful producer, he says, is no different from being a company CEO: "You have to create conditions where people can be innovative and can flourish in a personally supportive environment." The first play he invested in was *Raisin in the Sun* in 1961, which was the first black play on Broadway. He also invested in some of the early Neil Simon plays. His longest-running venture as an investor and producer is the award-winning *Stomp*, a celebration of rhythm in sound and movement and in comedy, which chalked up its 10th anniversary of continuous performances in New York's Orpheum Theater in February of 2004. Morty's association with *Stomp* is quirkily appropriate: rhythms have played a key role in his life in recent years.

Throughout his career as a successful businessman and producer, Morty was dogged with chronic ill health. Nothing life threatening, but doctors were unable to help. As a result, he decided to try Dardik's cyclic exercise protocol in 1991 and follows the program to this day. Here is how Morty describes that path:

"When I got out of college, I went through a period of real deterioration. I ate too much, drank too much, went out too much, didn't exercise. That was through the age of 35. At one point I was 210 pounds and feeling really out of shape. One day I was breathing so hard when I bent down to tie my shoes, and I said to myself, 'I have to do something about this.' I started to run; that was when running was coming into vogue. I ran myself back to health. That was in my 40s. When I was 50, to celebrate my birthday, I ran the New York marathon. I didn't finish it. At the 20-mile mark I broke down, hit the wall. That surprised me because I was very athletic.

"I had been feeling lousy on and off for years. Finally, after the marathon disappointment, I went to my doctor, who told me that my blood profile suggested I had a liver problem, so he sent me off to a liver specialist. The guy said it looked like a form of hepatitis, but he couldn't pin it down to a type, A, B, or C. He said, 'Your liver has probably 15 more years in it, no more,' which was startling to hear. That was 25 years ago, and I keep being told that my liver has another 15 years in it! I realized that being old is 15 years older than you are, no matter how old you are.

"Early in 1991 I picked up a copy of *New York* magazine and read an article about Irv Dardik's cyclic exercise program and the dramatic results it had achieved with people suffering from all kinds of conditions. I instantly felt I had an affinity for what he was talking about, a gut feeling that this is what I needed. I said, 'I'm going to track this guy down,' which I did after about a month. At first I talked to his wife and told her I was interested in doing the program because of my liver condition. About a week later Irv called me up and said he would do it. He hadn't worked with anyone with liver problems, and that intrigued him.

"We finally got together in the spring of 1992. He came to this apartment, and he spent half a day talking about his theory and his program, how it worked. After that he came three times a week, in the mornings, and I did weights and ran up and down the staircase here. He checked my heart rate, looked at how fast it accelerated and how fast it recovered when I rested. He worked with a yellow pad and figured out what I needed to do next. It was pretty unsophisticated back then, compared with the computerized analysis he uses now.

"After a month or so doing this, he sent me down to Florida for two weeks to work with Jim Loehr, his colleague. I was a tennis player—still am—and that's what Jim was doing as a form of cyclic exercise. Tennis is a great sport for this kind of exercise. I would do exercises like volleying at the net, overheads, to get the heart rate up. Then I would rest. Then more exercises. And so on. I can incorporate this regimen into a real game by sitting down between each point, to get my heart rate back down to 80, which would happen quickly. No one has to wait for very long; about 30 seconds between each point.

"I wish I could remember how soon it was before I began to feel any positive effects, but it wasn't very long, a month, two months maybe. I no longer felt blah. I felt energized. Pretty soon I was playing tennis with more energy. I was skiing more efficiently, doing black diamonds. I still do that, though I avoid the big bumps now! Then, early in 1994, I took myself off to my doctor because I had some rectal bleeding. He did a colonoscopy and found a tumor. I saw it on the screen, and it looked pretty simple. I said, 'Why don't you take it out right now?' He said, 'Not so fast, Morty, we have to do a biopsy.' It turned out to be malignant. I was pissed at Dardik. I called him and said, 'How the hell did this happen? I've been doing cycles for two years, and now this pops up!'

"When the surgeons went in there, they found that the tumor was really big and had grown through the wall of the intestine. They said that it must have been there for quite a long time, years. But the astonishing thing was that it hadn't metastasized. The doctors were shocked because normally they would expect it to have gone to the lymph nodes and the liver, given how big the tumor was and how long it had been there. I asked them how that could be. 'Oh, it must just be your nature,' they said, but Irv thought the cycles might have prevented it spreading. They put me on chemo for a year, as a precaution. I went up to Sloan-Kettering every Wednesday, and the amazing thing was that I never had any side effects—no nausea, no hair loss, the white cell level was fine, the red cell level was fine. I never missed a day of tennis; never missed a day of skiing. Again I asked the doctors, 'How could that be?' Again they said, 'Oh, it must be your nature.' I told them, 'I am doing these cycles. They must be doing something.' But the doctors were very dismissive. They couldn't see how doing these simple exercises a few times a week could have such a dramatic effect.

"I'm convinced the cycles did stop the tumor spreading, did forestall negative effects from the chemo. Who knows? But I do know that the worst thing that can happen when you do the cycles is that you feel terrific!"

Within a year of the publication of the *New York* article, Dardik and Godfrey were working with 50 or 60 clients, many of whom were paying good money. The tsunami-like response to the magazine article thrust Dardik into a sustained period of concentrating almost exclusively on the health aspects of his theory, although he continued to work on a paper that he hoped would be published in the scientific press. "All of a sudden, we were no longer in financial crisis, and we were able to begin paying off debts," says Godfrey. "But a lot of the money was used to hire exercise physiologists and exercise therapists that we trained in the program. At one point we had as many as 10 of them. Everything was great!"

A number of prominent businesspeople were among the throng of respondents, including publisher Si Newhouse, whose second son, Wynn, was wheelchair-bound with advanced multiple sclerosis. Newhouse paid a substantial sum for his son to be on the program, which enabled many others with limited resources also to participate.

In his continued voracious reading, Dardik kept coming across magazine and journal articles that pointed out that chronic diseases were much less common at the equator, sometimes dramatically so. For instance, multiple sclerosis is 80 times more prevalent in northern latitudes than at the equator. The incidence of breast, colon, and prostate cancer is also greater at higher latitudes; the same with diabetes and schizophrenia. "There's no explanation in traditional medicine for this anomaly," says Dardik, "but we reasoned that the wave patterns must be stronger at the equator, which is at the peak of the Earth wave. We therefore started taking people down to the Caribbean."

The first island they chose was Aruba, and among the people in the group was Dardik's daughter from his first marriage. She suffered from Crohn's disease, inflammation of the intestine that some think is caused by an autoimmune reaction, perhaps triggered by a viral infection. Dardik's daughter had particularly severe symptoms, including pain in the intestine and intestinal problems. By the time she was 18, she had extensive surgery, removing part of the colon and small intestine. A decade later, in 1993, shortly after she was married, she suffered another acute attack, this time including severe arthritis. She could barely walk and could only ingest liquids. Further surgery was planned.

"I said I would try to help her, take her to Aruba with us," says Dardik. "When we were at the airport, we had to help her with everything because she was so incapacitated. Within a week, she was running on the sand, her arthritis almost gone. The pain from the intestinal irritation was subsiding, and she was beginning to eat solid food. She was still taking steroids, but at a much lower dose. The combination of the exercise program and the drugs was working amazingly. The doctors told her that she couldn't have children, and now she's on her fourth!"

When they first started going to the islands, Dardik, Godfrey, and their clients stayed in hotels, but that proved to be far from ideal. "It wasn't like being at the farm, where there was just us, in nature," explains Godfrey. "We just didn't fit in. We would be going to bed early, while everyone else was up partying, screaming and yelling. And we would be up at sunrise, running on the beach, screaming and yelling. No one was happy."

Eventually, after trying several island locations, Godfrey found the perfect spot, a large beach house in western Barbados. "It had seven bedrooms and bathrooms, a pool, a separate cottage, and its own private beach," describes Godfrey. "There were white coral outcrops on either side of the beach, going into the sea. Anyone who wanted to walk on our beach would have to wade out into the sea to get to it. Most people didn't do that. It was just like being on the farm, in nature, but in a fabulous setting. We got up at sunrise, did cycles. More cycles in the late morning, and then again in the afternoon. We know now that that's not optimal, but we were still learning. People would take naps in the afternoon. In the evenings, we used only candlelight, and people went to bed very early. Everybody loved it."

One curious phenomenon that sometimes happens when people are at the peak of their exercise wave, Dardik and Godfrey observed, was that they would get extremely emotional, and this happened several times during the two-month sojourn on Barbados. "Some people broke down sobbing, others got very angry and would rant at whoever was working with them," recalls Godfrey, "and that was often Irv. We think that what happens in these cases is that the peak wave patterns tap into a similar wave pattern of a past unresolved emotional trauma, locked away. The peak of the waves stirs submerged waves, and they come flooding out. When someone is scared or angry, the physiological response, the waves, are very similar to intense exercise—heart pounding, blood pressure rising. Most people had no idea what the emotional reaction was all about while they were cycling or ranting. The *memory* of the past emotional trauma didn't come back, but the *feeling*, the physiological experience, did.

"I remember one woman, Anne, she was on the trampoline in the barn, and all of a sudden I heard this tremendous outburst of crying, and through her tears she started talking about a time when she was a kid, something dreadful happening. I don't know what it was. She was quite confused by the experience. But it didn't happen again. That was unusual, to have it resolved like that. Another woman, who worked for CNN, sometimes got very angry at the peak of the waves, but she had no idea why. She was in therapy at the time, and she and her therapist uncovered some stuff from her childhood, and when she came back to do cycles, the anger was gone.

"We used to warn people that something like this might happen, that the waves they are creating go deep. It is obviously not just exertion and the physiological response to it that triggers the outburst, because these people didn't experience them when they were exercising in a regular way. There's something fundamental going on when people generate powerful waves while at the peak of their cycles. It's a cogent mind-body connection, reconnecting your body to past experience at another level."

Many of the half dozen clients at the Barbados house had cancer of one form or another: kidney cancer, leukemia, breast cancer. "They were all doing extremely well," says Dardik. "The person with leukemia, his white cell count was going down. The woman with breast cancer, her tumor was shrinking. And this was after just a short period of time. After our time in Barbados, the guy with renal carcinoma went back home to an appointment at the Mayo Clinic, where they wanted to try an experimental treatment. I went with him. They did all the scans and said, 'Your tumors are shrinking. Some metastases, in the lungs and liver, have disappeared. So why would we make a patient out of you, with a new drug? Whatever you are doing, keep doing it.' He told them he planned to."

Life could not have been better for Dardik and Godfrey, difficult though their work was. They had the satisfaction of seeing clients respond positively, often dramatically, to the exercise program. Money was no longer the problem it had been for so long. Debts were being paid off. They had the luxury of traveling to beautiful places to do their work. Moreover, Dardik finally completed a paper on his theory. Called "The Great Law of the Universe," it was published a year later, in the 1994 spring issue of *Cycles*, the journal of the Foundation for the Study of Cycles. The foundation was established in 1941 to "foster, promote, coordinate, conduct, and publish scientific research . . . in rhythmic fluctuations in natural and social phenomena." His radical theory was to be out in the public arena. Dardik concluded his paper as follows: "Nature is wavenergy. Recognition of the Law enables man to create a true new beginning, a new way to reconnect and live among themselves and nature by developing a new, creative civilization. As the guiding principle of the universe, the Law instructs man in the way to use information and knowledge he has gained to address simple and complex issues the way nature has and always will."

In September 1993, Dardik and Godfrey were in the midst of planning a visit to Costa Rica, taking with them some of the people who had been at the Barbados house. The clients had rented the house themselves, so Dardik and Godfrey were not involved in those finances. They were simply going to turn up and work with the clients for two months. And then "the letter" came.

8

The Trough

"Men build too many walls and not enough bridges."

—SIR ISAAC NEWTON

ELLEN BURSTEIN WAS A CONSUMER ADVOCATE, a "fraud buster" for a television channel in Orlando, Florida. In May 1986, Labe Scheinberg, a neurologist at Albert Einstein College of Medicine in New York City, broke the news to Burstein that she had multiple sclerosis, thus solving a three-year-long mystery of why her left leg kept collapsing under her while she was running. Burstein, a self-confessed type-A personality, as MS sufferers often are, was determined to beat the affliction and eager go back to her passion, running. The disease progressed rapidly, however, and within four years she had great difficulty walking. She used a walker or a wide-based cane to get around and sought the support of a wall while walking in her office. While on assignment, she got around in a three-wheeled motorized scooter.

In the spring of 1991, a friend gave Burstein a copy of Tony Schwartz's *New York* magazine article about Dardik and his exercise program. Ever on the lookout for charlatans because of her line of work, Burstein was nevertheless sufficiently convinced that Dardik's claims were genuine that she ferreted out his unlisted phone number and called his office. She was told there was a waiting list of 2,000 chronically ill people, as a result of the article. Burstein is nothing if not persuasive, and Dardik eventually agreed to take her on, for a fee of $100,000.

On June 10 she took a leave of absence from her job and moved to New York City, following Dardik's recommendation, to stay with her younger sister, Jessica. On June 11, the day before she was due to meet Irv Dardik for the first time, her physician told her that, given her rapid deterioration, she would probably be a quadriplegic within a few years. At best, he said, she would spend the rest of her life in a wheelchair or on a scooter. "I am furious with Labe," Burstein later wrote in her book, *Legwork*. "He doesn't understand that I won't compromise. I refuse to settle for mobility in a scooter."

At his first meeting with Burstein, Dardik agreed to start her on the program within a few days. He also introduced her to Michelle Morelli Weiss, an exercise therapist he had hired as a personal trainer for his clients. Burstein continued to take the drug Cylert, to control fatigue, as she began the program. "But by early August, after just two months of exercise cycles, I have so much energy that I am able to cut the dosage in half," she wrote. "I am gaining strength; my feet are no longer numb; I stopped using a brace on my left foot; and for the first time since the MS diagnosis in 1986, I can feel the floor beneath my bare feet." That month, Burstein moved to her family home on Long Island, and three months later moved back to Orlando. Morelli Weiss and Dardik made occasional visits to go through cycles with her. At other times they did it over the phone or arranged for a local therapist to be with her.

Burstein continued to make slow progress. And early in January, during one of Morelli Weiss's visits, she was able to get her heart rate up to 160 two days in a row and walk the entire length of her house. The miracle she had been hoping for appeared to have happened. She was so thrilled that she paid a visit to her former office mates, walking unaided for the first time in years, and was greeted by a standing ovation. All this was reported in a short *Orlando Sentinel* article titled "Back on Her Feet: Unconventional Therapy Helps TV Consumer Reporter with Multiple Sclerosis Return to Action." The article quotes Burstein's neurologist as saying the previous August, "There is no question that she has shown slow but definite improvement, graphically so as compared with my earlier assessment in April of this year." He recommended that she continue working with Dardik.

The "miracle" was, however, a onetime event. Burstein never walked like that again, and her progress trailed off. She became less and less satisfied with Dardik's work with her and began to harbor suspicions that she had been duped. In August, she agreed to spend a few days at the farm, even though she had by that time concluded that Dardik was a fraud. She returned home with the intent, as she wrote, of "putting Irv out of business."

To that end, she hired a private detective, Alice Byrne, a friend of Burstein's other sister, Patricia. Byrne, who happened to have multiple sclerosis, approached Dardik as a prospective client. Dardik was at first reluctant to take her on but eventually agreed. Months passed, but in July 1993 Byrne finally had a consultation with Dardik at the farm. She took two people with her, whom she described as a nephew and a friend. In fact, they were employees of a major television network, and they filmed scenes and conversations in and around the farm. (A problem with the battery caused the loss of audio, so it could not be used.) During that meeting, Dardik agreed to take Byrne on as a client, and a few days later faxed her a contract that included a statement to the effect that the program was still experimental and, therefore, no particular desired outcome could be guaranteed. Byrne had no further contact with Dardik.

Meanwhile, Burstein had been busy on her own account. During her short stay at the farm the previous August, she says, she found a list of Dardik's clients and their phone numbers in her cottage and copied them onto her computer. "It's news to me that the list would have been in the cottage," Godfrey now says. "As far as I knew, we only kept it in the office in the farmhouse." In any case, Burstein called people on the list, saying that Dardik was a fraud, and urged them to join her in an action against him. She also filed a complaint to the New York Department of Health on December 22, 1992, saying that she believed that Dardik was "guilty of medical incompetence, negligence, and fraud."

The wheels of the medical establishment were now set in motion against Dardik. They moved slowly, however, and it wasn't until the following September that "the letter" arrived at the farm, in the form of a

Notice of Hearing from the New York Department of Health, State Board for Professional Medical Conduct. Two key elements of the complaint were, first, that Dardik had promised to cure Burstein of her multiple sclerosis, and that he had failed to do so; and second, that he had promised her his constant personal attention throughout the program, and that he had failed to do that, too. The charges also included the assertion that Dardik's statements about the efficacy of his program were "false and Respondent knew they were false."

"The suggestion that what I claim for the program is false is ludicrous," says Dardik. "I have seen so many sick people benefit from the program. I was then, and I still am, absolutely confident that the cause of chronic diseases is a disrupted wave patterns in individuals, and that the cyclic exercise protocol can reverse these disorders by re-normalizing the wave pattern." As for the charge that Dardik "knew [his claims] were false," Dardik had this to say at the hearings: "At the very beginning of the program, I knew she was one of those consumer advocates. If I truly was a fraud, I would be the last person in the world to do this program with someone like her."

The alleged promise of a cure was, however, central to the charges. "I did not use the word *cure*," contends Dardik. "You can call what I do a cure, but I prefer to call it reversing disease. What I do is help people connect with their bodies in a different way, through making healthy waves. It's a different way of thinking about health. It is *you* taking charge of your own health."

Godfrey says the same thing. "Irv never uses the word *cure* because it isn't consistent with the theory," she says. "You don't *cure*. You change a wave pattern *to cause health*. There is no curing in the way the medical establishment thinks about curing. If you reinduce unhealthy waves, by not doing the cycles or pushing yourself too hard physically or emotionally, disease symptoms will reappear. That's what happened with Ellen. She took a break from the cycles in March, a very bad time of the year to do that, before spring comes. As a result, she crashed. We warned her that that might happen, doing what she was doing."

Nancy Kaplan, who also had multiple sclerosis and was included in the *New York* magazine article, says that when she approached Dardik, "he never used the word *cure*. He just told me that he would help me get past this. And he did." Nicholas Hall, who spent a lot of time with Dardik and his clients, corroborates Kaplan's statement. "Never once did I hear Irv claim that this was a cure for anything," he told me. "What I heard was Irv talking about the theory and how making healthy waves could help reverse disease."

In any case, before Burstein began the program, her mother, a former State Supreme Court judge in New York, and her brother, a lawyer, worked with Dardik's lawyer to draft a contract that stipulated that the fee would be refunded if, for any reason, Dardik was unable to complete the program. It also contained a paragraph that stated explicitly that the program did not guarantee a cure and that the program was experimental.

As for the claim that Dardik promised full-time, one-on-one attention, Godfrey is dismissive of the notion. "That's ludicrous," she now says. "If you think about it, here is this man with a full roster of clients and a crew of physical therapists; would he promise his full attention to just one client? That makes no sense. We promised to make available an exercise therapist, Michelle, to be with her when she did the cycles, especially at the beginning, and Irv would supervise the program. Irv would also be with her from time to time. That's why we suggested she move to Manhattan from Florida, which would be more convenient. Our understanding was that she would be in Manhattan throughout the program. When she moved to Long Island, that made it more difficult for Irv to go frequently, but we managed. Orlando was a different story, and we had to hire local therapists to help out when Irv or Michelle couldn't be there. By moving from Manhattan, Ellen broke the understanding we'd had at the beginning."

These points, and many more, would be raised at the medical board hearings scheduled to begin in November 1994. Meanwhile, Burstein's actions were having more immediate, detrimental effects on the Dardik household. "Ellen is a very compelling, very articulate, bright woman, who had enormous credibility," says Godfrey. "When she called the clients on the list, some with the same personalities, the same disorder, the same home-life problems, they listened. She explained what course of action she was planning and asked if they would be co-plaintiffs. Out of the whole roster she got just three to sign with her, other MS people, even though they had no unique complaints of their own. Alice Byrne also joined as a plaintiff, which was a little bizarre, as she was a private detective looking for evidence to sink Irv and never had any intention of being on the program. The other clients Ellen talked to got nervous. They didn't know what they were about to be dragged into, so many of them pulled out of the program, even though they were being helped."

By the end of 1993, income had virtually dried up. Foreclosure on the house once again loomed. Dardik and Godfrey would have been in truly dire straits but for financial support from several prominent, wealthy clients whose faith in the program and in Dardik's integrity remained solid. One of these "angels" was Dick Fox, founder and chairman of Fox Companies in Philadelphia and for 15 years chairman of the board of trustees of Temple University. "I first met Irv 15 years ago," Fox told me. "I was commuting back and forth to London once a month and feeling lousy. I was on my way to becoming ill. The first time I was with Irv, I couldn't get my heart rate above 120, so I began working with him on the program. Within a few months, my whole life was turned around. I felt completely healthy for the first time in years, and I have Irv to thank for that. He has a very special insight into health and disease, and I could see that the medical board was a travesty, and, as a friend, I just wanted to help."

Godfrey, meanwhile, threw herself into a new role, that of legal assistant. "I spent time getting papers together," she recalls, "doing research, writing in preparation for the case, doing whatever I could do to help our lawyers, to try to keep the costs down. Obviously, throughout all this, we couldn't take on any new clients." The stress of all that unfamiliar work, the return of financial crisis, and uncertainty about the outcome of the impending hearings took a tremendous emotional toll on Godfrey. "My work on the case made me responsible for what was going to happen on the case," remembers Godfrey. "Irv was completely emotionally unavailable and withdrawn. He was doing what he needed to do. So I was in one of those phases, doing anything that needed to be done to support him. That rendered me pretty useless from time to time, and I would retreat into an upstairs closet, curled into a fetal position. When you have little kids, as I did, they are like hunting dogs; they find you and drag you out. That happened many times. The closet was my hiding place, under the clothes, back in the womb. It was a truly dreadful time, the dark night of the soul."

By contrast, Dardik's lawyers were confident and reassuring. As there had been no claim of injury, serious sanction was unlikely, they told him. The worst that could happen would be a small fine.

The medical board hearings began on November 15, 1994, and continued until the beginning of February. The board met on 10 occasions, usually for a few hours at a time, at the offices of the New York State Department of Health, near Penn Station in Manhattan. Three members of the State Board for Professional Medical Conduct and an administrative law judge sat at the blunt end of a U-shaped table in an ordinary-looking conference room; Dardik and his lawyer, Eugene Scheiman, sat on one side of the U. Boxes of heart rate graphs of Dardik's clients were arranged on a side table but never actually referred to. Burstein, three other of Dardik's multiple sclerosis clients, and Alice Byrne were called as witnesses, as were Burstein's mother and sister Jessica, as well as Alan Tuchman, a specialist in multiple sclerosis from New York Medical College, who told the board that "there is no cure for multiple sclerosis." Dardik was limited to a few witnesses, including Godfrey, Nancy Kaplan, and Nicholas Hall. "The ambience was one of stern disapproval, of admonition, right from the beginning," remembers Dardik. "It felt as if the whole thing was a foregone conclusion, and they were going through the motions."

It's interesting to note the intellectual context in which all this took place. "This was the height of anti-alternative-medicine sentiment," explains Dardik, "and I was the epitome of alternative medicine to the board members." For instance, shortly after Dardik's encounter with the medical board, the New York Academy of Sciences convened a three-day meeting of scientists and medical people, called "Flight from Science and Reason." Malcolm Browne, a reporter for the *New York Times*, wrote, "Defenders of scientific methodology were urged to counterattack against [alternative medicine]." In his article, Browne quoted Gerald Weissmann of New York Medical Center as saying, "Medicine and science today are being confronted by lunatics, fascists, and the practitioners of bizarre magic."

How things change. In the decade since that meeting, the National Institutes of Health established a separate division, the National Center for Complementary and Alternative Medicine, with a budget of more than $120 million in the year 2005. Alternative medicine is now mainstream. But during the time of Dardik's hearing, there was only one way to do medicine: "You followed the rules, and heaven help you if you didn't," explains Dardik. "And I wasn't following the rules, never have. Here I was saying that, from the perspective of SuperWave theory, chronic disease can result from a flattening of the waves, with a cascade of ill effects throughout a person's physiological systems. And I was saying that people can take charge of their own health by restoring healthy waves, through cyclic exercise, and so reverse the disease. But in traditional medicine, a person has a disease, and it is the role of the physician to go in and cure that disease, using the tools of traditional medicine. I was essentially threatening their belief system. Because of that clash of perspectives, I felt the board was strongly biased against me all along."

Biased or not, the board did not look favorably on Dardik. A summary of the charges, as set out in the board's papers, is as follows: "[The board] charges the Respondent with professional misconduct by fraudulently practicing medicine, by exercising undue influence and exploiting his patients for his own financial gain, by guaranteeing a cure to patients, by revealing personal identifiable facts, data, or information in a professional capacity without prior consent of the patient, and by engaging in conduct evidencing moral unfitness to practice medicine."

On March 24, the board issued its findings, stating that Dardik was guilty of fraud, of exercising undue influence over his clients, and of guaranteeing a cure. It noted that the contract that had been drawn up, which stated that a cure was not guaranteed, contradicted what Burstein alleged (and Dardik denied under oath) he told her to her face. Dardik was found not guilty of revealing personal information and not guilty of being morally unfit to practice medicine. The board voted unanimously to revoke Dardik's license to practice medicine in the State of New York.

Dardik's lawyer, who was an expert in medical malpractice, was aghast, and he immediately filed an appeal to have the decision overturned or, if it was sustained, at least have the penalty overturned. Burstein's lawyer was equally shocked, believing that the decision was too lenient. He urged the Review Committee to find Dardik "guilty of moral unfitness to practice medicine" and have Dardik fined $100,000 "to both punish the Respondent for his misconduct and to send a message to the public that conduct such as the Respondent's is not acceptable." On July 28, the Review Committee issued its findings, in which it upheld the board's findings of guilt and the penalty of license revocation; it declined to find Dardik morally unfit to practice medicine, but it did impose a fine of $40,000, representing $10,000 for each of four of the plaintiffs but excluding the fifth, Alice Byrne, the private detective. Dardik also faced $250,000 in legal fees.

Shortly after the review board submitted its decision, NBC television's *Dateline* program did a piece about the affair. Dardik declined to be part of it, but Burstein participated, as did Labe Scheinberg, her neurologist from Albert Einstein Medical Center. He was very dismissive of Dardik's theory, saying, "The day Dardik is proved right is the day pigs will fly." That remark prompted Dardik's friends, relatives, and faithful clients to send him renditions of pigs flying. Now, every room in the Dardik household has winged flying pigs in every imaginable form—statuettes, large statues, candles, painted tiles. "They are getting ready to fly," jokes Dardik.

At first, Dardik was crushed by the loss of his license, and, typical of him, threw himself into reading, this time the history of science. "I kept coming across instances in science and in medicine of people who were ridiculed for their ideas but were ultimately proved right," he remembers. There was Ignaz Semmelweis, the Hungarian doctor who in the mid-19th century discovered that puerperal fever, also known as childbirth fever, was contagious, and that its incidence could be drastically reduced if doctors washed their hands between seeing patients. He published his ideas in 1861, was lambasted for them, suffered a nervous breakdown, and was committed to an insane asylum in 1865, where he soon died of blood poisoning. In the 17th century, William Harvey proclaimed that the heart was in fact a pump that distributed blood throughout the body. He based his theory on many dissections of animals. The received wisdom of the time was that the heart was the seat of the soul and that the pounding in the chest was but the soul speaking to us. Harvey's idea was laughed at for years but, of course, was eventually proved correct.

One of the reasons that Copernicus's suggestion that the Earth moved around the sun was so fervently rejected, apart from controverting the teachings of the church, was based in language. In *Structure of Scientific Revolutions*, Thomas Kuhn notes that those who criticized Copernicus did so because part of what the word *earth* meant was "fixed position." "Their Earth, at least, could not be moved," by definition, writes Kuhn. "When I read this," says Dardik, "I thought of the medical board, who told me that 'chronic diseases are incurable, so how can you, Dr. Dardik, cure the incurable?' I, of course, wouldn't use the word *cure*. I would say 'reverse the disorder by creating healthy wave patterns,' but the board couldn't, or wouldn't, hear that."

One of Dardik's favorite passages from this exploration of the history of science comes from a letter Galileo wrote to his friend Johannes Kepler in 1630. Galileo complained about those who rejected his ideas but refused to look through a telescope so they could see what he was seeing: "My dear Kepler, what do you say of the leading philosophers here to whom I have offered a thousand times of my own accord to show my studies, but who, with the lazy obstinacy of a serpent who has eaten his fill, have never consented to look at the planets or moon or telescope? Verily, just as serpents close their ears, so do men close their eyes to the light of truth."

"When I found this, I thought of the medical board," says Dardik. "I wrote in the margin next to it, 'They think they own the truth, and aren't prepared to look at anything that is not *their* truth.' I know I am right, and one day I will be proved right."

Dardik found himself comforted by these and other such examples of new ideas at first being rejected and then vindicated. What had initially been the pain of losing his license became, he says, "a badge of honor, for going against the system, and I'm proud of that!"

Godfrey, meanwhile, was in the closet, under the clothes, wondering where the money was coming from to pay the bills.

Just when it looked as if things couldn't get worse, they did. It was seven in the morning on a late August day, just before Labor Day weekend, not long after the review board's decision. The Dardik household was surprised by loud knocking on the front door—and even more surprised when Dardik opened the door to reveal three uniformed policemen standing there, with three police cruisers idling in the driveway. "Dr. Dardik," one of them said, "you are under arrest!" Everyone in the house was shocked. "My 96-year-old grandmother had just moved in with us the day before," recalls Godfrey. "She is standing there in her pajamas. The kids are in their pajamas, scared as hell. We didn't know what was happening. They just took Irv away in handcuffs, and they wouldn't tell me what it was about. They wouldn't tell me where they were taking him. They just kept saying, 'You will hear from your husband later.' I was frantic."

The reason for the arrest was for Dardik's nonpayment of alimony and child support. Under the terms of the 1984 divorce, Dardik was ordered to pay $150,000 a year, which was later reduced to $90,000. He'd had a ton of money while practicing surgery, and that was supplemented handsomely by royalties from the biograft; hence, the generous settlement. But he stopped doing surgery in 1980, and the flow of royalties dried up not long after the divorce. Without any serious income, Dardik fell behind in payments, so that by that August day, he owed $850,000. He had managed to make payments from time to time on an agreed schedule, but for some time prior to the arrest, he had made no payments. According to a standing bench warrant, Dardik would go to jail if he did not meet the schedule, and his ex-wife decided to activate that option.

"Hours went by, and I was beside myself," recalls Godfrey. "I made dozens of phone calls, trying to find out where Irv was, trying to get a lawyer, but that was impossible just before the holiday. Media helicopters were buzzing around the farm, taking pictures. I had no idea what was going on. Finally, Irv calls and tells me he is in court, and that he has the name of a lawyer. I called the guy and made arrangements to meet Irv in prison, so I could take him to the lawyer." Godfrey immediately drove to Bergen County Jail in Hackensack. "When we tried to drive off to see the lawyer, there were reporters everywhere, sticking their cameras into the car," says Godfrey. "I thought, 'We've got to get out of here,' so I slammed down on the accelerator and sped off. Unfortunately, it was into a dead end, so I had to turn around and run the gauntlet again! It was horrible."

The jail was a kind of halfway house, where some of the inmates were allowed to go out to work during the day and return in the evening. "There weren't any individual cells," says Dardik. "It was a big open space with bars around it, with 50, maybe 60 guys in there. Initially, they put me in the back, but then one of the guards said to me, 'We can't leave you there. You could be dead in the morning because you're a deadbeat dad.'" That same phrase was used in the next day's report of the incident in the local paper, *The Record*, except that it was enhanced somewhat to "County's Worst Deadbeat Dad." That characterization was a little harsh because, as Dardik's brother, Herbert, noted in the newspaper, three of the four children "are married and have children, and the fourth is at university and doing very well on his own."

Dardik was allowed out each day to look for work so he could make payments toward the $24,000 that was immediately owed. In fact, he went home to the farm, and continued his reading of science and thinking about the theory. "I took heart rate monitors back to the prison with me," says Dardik, "and pretty soon I had a lot of the guys doing cycles. They had universal machines, so that was easy to do. They guys really loved it, and the guards got interested too." Dardik was in jail for just a week, and by the end of that time he had a coterie of inmates enthusiastically working their way through the cycles protocol. "Irv is irrepressible," says Godfrey. "He's unsinkable. Whereas most people would be depressed by the situation, Irv just saw it as another opportunity, a way to see what it was like in prison, to see how we cage people, and to see the effect on their spirits when they were doing the cycles."

9

Body and Heart Waves

"The atoms come into my brain, dance a dance, then go out; always new atoms but always doing the same dance, remembering what the dance was yesterday."

—RICHARD FEYNMAN

DARDIK WASN'T EXACTLY BROODING over the medical board's refusal to take seriously his idea that an individual, by making waves through cyclic exercise, could take charge of his or her own physiology and so enhance health. He was disappointed, yes, but not really surprised; the notion was so outside the realm of traditional views of physiology and medicine that the board members simply weren't able to hear what was being said. What he needed was some form of independent corroboration from the world of medicine, but he had no inkling of what that might be or from where it would come.

Then, in September 1995, just two months after the review board had upheld the medical board's decision of "guilty" and increased the penalty, an envelope arrived in the mail, containing a photocopy of an article published a year earlier in the journal *Circulation*. The article was titled, "Reduced Heart Rate Variability and Mortality Risk in an Elderly Cohort: The Framingham Heart Study." The sender of the article, a client of Dardik's for five years, appended a simple note, which read: "Irv, I think this is something like what you are talking about. Maria."

"I scanned the article, and I knew that Maria was right," recalls Dardik. "This was the first confirmation from the medical literature that the cyclic exercise program was tapping into something fundamental, and it explained why it works. I quickly called Maria and said, 'Thank you, thank you, thank you. This article tells me to keep going.' She said, 'Yes, Irv, keep going, keep going!' I was running around the house out of sheer joy."

The heart rate variability that the authors of the article talked about is a simple insight into the way the heart works and its association with health. When a nurse takes your pulse at the doctor's office, he or she will give you a number—60 beats a minute, for example. But that in fact is just an *average*. If you were to plot each beat throughout that minute, you would find that it is not metronomic, with one beat every second. Instead, you would see that the time between each beat, the interbeat interval, varies: at some point during that measured minute, your heart might be beating at an average of 70 beats a minute, say; and at another point, it might be an average of 50. In fact, every time you breathe in, your heart rate increases a beat or two, and it falls when you breathe out. Heart rate variability, then, is a measure of how much the interbeat interval varies between consecutive heartbeats when you are at rest.

The authors of the article had surveyed 736 elderly people over a period of 4 years and found that those with high heart rate variability lived considerably longer than those with low variability. "To many people, including a lot of people in medicine, this seems counterintuitive," explains Dardik. "Traditionally, doctors thought of a healthy physiology as being one that is finely controlled; that order was good, and disorder was bad. In fact, we now know that the opposite is true."

Researchers who study nonlinear dynamics in physiological systems, such as Ary Goldberger of Harvard Medical School in Boston, love to play tricks on audiences, professionals and the interested public alike. He puts up two traces of the heartbeat of real patients, one of which is nice and regular, the other distinctly irregular, not in interbeat interval but other measures of order and disorder, or complexity. "One of these patients died of cardiac arrest soon after these traces were taken," Goldberger would tell his audience. "Which one was it?" Inevitably, the majority of people opt for the more complex beat. "Wrong," Goldberger would reply. "It's the one that looks reassuringly regular, or ordered. The one that is relatively disordered, or chaotic, is the healthy patient. Obviously, too much chaos is a bad thing, but a moderate amount allows the heart to respond to sudden changes. A complex heart rate is a healthy heart rate." It's a bit like the tennis player who is about to receive a fast service. She doesn't just stand there, immobile, waiting for the ball to be let loose. Instead she skips around on her toes, moving this way and that a little, ready to respond to the unexpected.

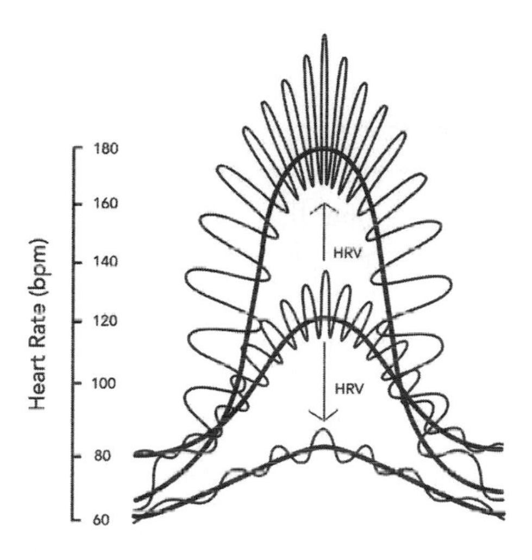

The HeartWave in Health and Disease. A healthy individual is able to produce a robust HeartWave (higher heart rate maximum, greater heart rate variability) when exercising and recovering; the HeartWave in people with chronic disease is much flatter (lower maximum heart rate, lower heart rate variability).

Dardik had contacted Goldberger a decade and a half earlier, in 1990, because he liked what Goldberger was doing with complexity and heart function and intuitively felt that there was some kind of connection between what they were both interested in. Goldberger was polite, but effectively told Dardik to "come back when you have something concrete to show me." The reported connection between heart rate variability and prediction of mortality in the *Circulation* paper was definitely in the right direction, Dardik believed, but he wasn't ready to make the return visit to

Boston just yet. He felt he first must delve deeper into the literature to see what kinds of patterns he would find, and he would need something concrete to show Goldberger.

What he found was a mystery, at least to traditional medicine. "In every chronic disease, whether it is diabetes, Parkinson's disease, multiple sclerosis, cardiovascular disease, cancer, even behavioral disorders, you name it, *heart rate variability is reduced*, and it reduces steadily with increase in age, too," explains Dardik. "Historically, medicine has looked for a specific cause, such as a molecular abnormality or a virus, to explain a specific disease. But the discovery of a *single* risk factor, lowered heart rate variability, tells me that there is some underlying phenomenon that connects everything to everything else that medicine has not understood. The SuperWave theory can explain it."

First, a little background: The maximum heart rate in a healthy individual is calculated as 220 minus that person's age. For a 20-year-old, the maximum heart rate is 200; for a 40-year old-individual, 180; and so on. The next thing to note is that the bigger the difference between *resting* heart rate and *maximum* heart rate in the HeartWave during exercise and recovery, the higher the heart rate variability. Higher variability is associated with a longer life and less susceptibility to chronic disease; lower variability is associated with a shorter life and greater susceptibility. Why should this be so?

Heart rate variability is the result of a dynamic dance between the two major branches of the autonomic nervous system, whose job it is to control the functions of virtually every organ and every physiological system of our bodies, and it goes on without us being aware of it. Specifically, the so-called sympathetic branch of the autonomic nervous system is associated with the fight-or-flight response, when heart rate, oxygen consumption, and blood pressure rise in preparation for urgent physical action. On the other hand, the parasympathetic branch of the autonomic nervous system governs the rest-and-digest response, or recovery from exertion. Traditionally, medicine describes these activities as being in dynamic balance, just as we did at the beginning of this paragraph. But in Dardik's view, they are in fact a wave continuum, just as exercise and recovery are a wave continuum of energy expenditure and energy recovery. We can see, then, that all the physiological systems of the body are in effect in communication with each other, mediated by the autonomic nervous system. And the more these systems oscillate, or make waves, the healthier they will be, just as Nobelist Ilya Prigogine said.

Let's put this graphically. Think of a small pond scattered with lily pads. Now imagine that the pond represents the human body, and each lily pad is some metabolic function, such as the immune defenses, liver function, digestion, or mood. And imagine that the surface of the water is in constant motion up and down, so that the lily pads are in constant motion, oscillating, moving up and down. That's a healthy body. In an unhealthy body, as in a stagnant pool, there would be much less motion of the water, much less up-and-down movement of the lily pads—much less oscillation of the physiological systems.

Now let's toss a rock into the middle of the pond and watch what happens. Ripples, or small, concentric waves, begin to spread throughout the pond, eventually reaching the banks, and then reflecting back again. The lily pads are in motion again, bobbing up and down, oscillating, as if they are talking to each other: nothing is untouched; everything is moving to the same rhythm; everything is in tune. That's a healthy body restored, with multiple waves waving to a ubiquitous rhythm.

In the context of the SuperWave theory, the "rock" in this illustration is repeated cycles of exercise and recovery, whose principle effect is to increase the range from resting to maximum heart rate, and therefore to increase heart rate variability. "Heart rate variability is a window onto *all* physiological functions," explains Dardik. "By grabbing hold of heart rate variability, so to speak, and pulling it up with the exercise program, you enhance the oscillation and health of all physiological systems, and you *cause health* in the body as a whole and reverse disease. In one stroke, the health efficacy of the cyclic exercise protocol is explained! When I realized this, I was reminded of the time a few months earlier, when I was trying to explain the HeartWave to the medical board, and they were looking at me like I'm a nut, that what I was saying couldn't possibly have anything to do with health. Wrong!"

Compared with cyclic exercise, *sustained* exercise of the sort promulgated by proponents of aerobics doesn't have the same effect and is actually detrimental. "Prolonged, continuous training in a narrow target heart rate zone leads to a lower heart rate variability," explains Dardik, "because you typically are not making repeated waves from *resting* to *maximum* heart rate, as happens with my cyclic exercise protocol. It's the repeated waving within that range that enhances that range. I knew from my experience at the Olympics that distance runners were always getting sick, with respiratory infections and so on, whereas sprinters weren't. The distance runners' immune defenses were obviously compromised, and now I can see why." Dardik wrote up his findings in a paper that was later published in the Spring/Summer 1997 issue of *Frontier Perspective*, the journal of The Center for Frontier Sciences, based at Temple University in Philadelphia.

Two weeks after Dardik received the heart rate variability paper in the mail, which validated his rationale for causing health through cyclic exercise, out popped another paper—this time from his fax machine—giving his ideas a further boost. The cover sheet from Dick Fox read, "Irv, I thought you would like to see this." The title of the article, from an exercise magazine, read, "Ken Cooper's Change of Heart." Cooper had launched the aerobics revolution with his blockbuster 1968 book, *Aerobics*, which sold more than 30 million copies worldwide and was translated into 41 languages. The message of that book, essentially, was "more is better," in terms of exercise and health benefits. By 1995, Cooper had changed his view. "His doubts began with the death of two friends, both of them elite athletes," the article began. The first friend to die, at the age of 50, was Werner Tersago, a world-class runner, who succumbed to a brain tumor in 1986. The second, Sy Mah, who held the world record for most marathons completed, died at age 60 in 1988, also from cancer.

The article explained that these events prompted Cooper to dive into the literature, looking for explanations of the fate of his friends. His conclusion: "Too much of a good thing may, in fact, be dangerous to one's health." A few months after the publication of this article, Cooper wrote one of his own, in *Health Confidential*, titled "Exercise Pioneer Dr. Kenneth H. Cooper Explains the Dangers of Too Much Exercise." "When I was writing my first fitness book . . . I believed that a regimen of sustained, strenuous exercise was necessary for health and longevity," he wrote. "But it turns out that serious exercisers may be unusually vulnerable to life-threatening illnesses. . . . In my practice, I've seen many cases of cancer and heart disease among elite athletes." He went on to say, "Very moderate physical activity gives the biggest health return on your investment of time and energy." That's a dramatic reversal of philosophy, by any account, and Dardik was thrilled to see Cooper, of all people, lining up with what he had been saying for so long.

"Cooper has all kinds of explanations for why sustained exercise is dangerous, such as the harmful effects of building up excessive free radicals during sustained, strenuous exertion," observes Dardik. "But the fundamental cause is that sustained exertion flattens the body's waves, including heart rate variability, causing disease." Dardik did a literature search of his own and came up with a slew of papers that confirmed Cooper's fears—and his own expectations. "I found paper after paper that showed the benefits of moderate exercise and the ill effects of heavy exercise, such as reduced immune function and higher incidence of chronic disease, including an increase in cardiac arrhythmias in elderly people with a lifelong history of regular strenuous exercise. I knew I was right all along, and the literature proves I was right!"

Gratifying though this double validation was for Dardik, he still had to find a way to pay the bills. Fortunately, two clients of considerable financial means contacted him around the end of 1996, both through friends of friends. One was a local industrialist whose son's asthma was so severe that he was frequently hospitalized. "I told the father that in order to have the best results as quickly as possible, I needed to take his son south," says Dardik. "He agreed, so I took the son to Club Med in Martinique, initially for a month. He made such good progress that I persuaded the father to extend the therapy for another month. The kid's asthma virtually disappeared, and he was able to get off most of his drugs."

The second client, a local man in the construction business, had suffered several strokes that left him paralyzed down his left side, and his speech was impaired. "We worked with him every day at his Florida home for a couple of months," says Godfrey. "We raised his maximum heart rate by 50 beats over that time. He became more and more mobile, and his speech started to come back, too. I remember one day toward the end of our time with him. He was walking unaided across the room when his nurse, a black Baptist woman, came down the stairs and saw him. She was so happy. She threw her arms in the air and shouted, 'Praise the Lord! Praise the Lord!' Geoffrey [not his real name] said, 'Yes, praise the Lord, but don't forget the doctor!' It was quite a scene."

This wonderful story has, however, a sad ending. Geoffrey returned to New Jersey and walked into his company's boardroom. He was chairman of the board, but during his incapacity, his son had taken over. "Everyone was shocked to see him walking, including the son," says Dardik. "Geoffrey sat down and started to run the meeting, which he did very well. Afterward, the son blew up because he was angry that he had been usurped, and he threatened to having nothing more with the business. Geoffrey was crushed. He called us and said he wasn't going to do the cycles anymore. He said, 'I have a choice between being well and not having my son, or not being well and having my son. I'd rather have my son.' He crashed, physically, and went back to the wheelchair."

Working with these and a few other clients not only gave Dardik and Godfrey some much needed income, but it also allowed them to refine the cycles program. At the time, Dardik was reading a lot about the lunar cycle and physiological correlates with it, and about fractals. The two converged in a way that completely reshaped the program. Dardik read that the lunar cycle is a wave of energy, with the full moon being the time of high energy, and the new moon, low energy. He also knew that testosterone levels peak at the full moon, which is also the time women in foraging societies ovulate. Tides peak at the full moon, as do tree diameters. "Alison and I were mulling this over, and we said, 'Look, what about daily cycles, the circadian rhythms. There's a peak in the late afternoon, between 3:00 and 6:00, when blood pressure, temperature, cognitive keenness, and physical-performance efficiency are at their highest. Why don't we do the cycles so that they track the energy wave during the day, and superimpose that on the lunar cycle? That way, the circadian rhythm is mapped onto the lunar rhythm, the one being a fractal of the other.' The idea emerged just like that, and it was one of those intuitive things that just *feels* right."

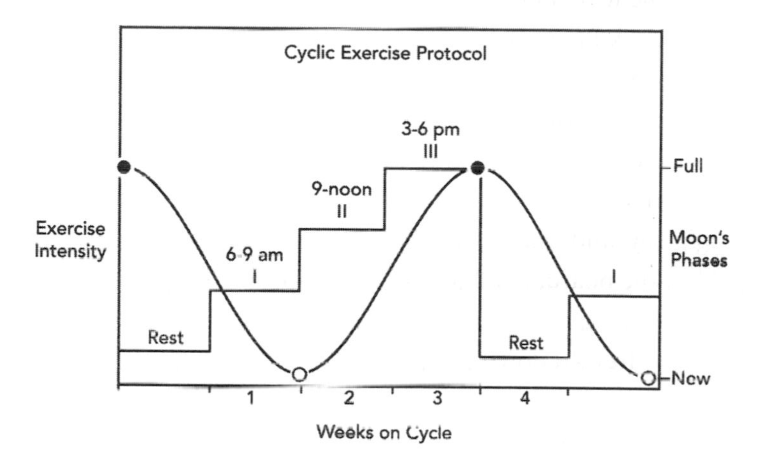

Cyclic Exercise Protocol. The intensity of exercise is tuned to the lunar energy cycle: Week I is lowest intensity, going into the new moon, done between 6 and 9 am; week II is higher intensity; done between 9 and noon; week III is highest intensity, going into the full moon, done between 3 and 6 pm; week four is a rest week.

The newly emerged protocol looks like this: In the week going into the new moon, the lowest energy segment of the lunar cycle, exercise cycles are done in the early morning, between 6:00 and 9:00, which is also close to the lowest energy segment of the day. The cycles themselves are also low energy, very gentle, maybe just three or four moderate bursts of exertion and recovery, three times a week; these cycles are designed to kick-start the metabolism, to help people pull out of the energy trough they've gone into during the night and early morning. The week after the new moon, a time of rising energy in the lunar cycle, exercise cycles are done in later in the day, between 9:00 and noon, a time of rising energy in the circadian rhythm. These cycles—again, three or four bursts of exertion and recovery, three times a week—are kicked up a notch in terms of intensity and heart rate targets. Finally, in the week leading to up the full moon, the time of maximum energy in the lunar cycle, the exercise cycles are done between 3:00 and 6:00, the time of maximum energy in the circadian cycle. Tracking this pattern, the exercise cycles are now done at maximum intensity, reaching the highest heart rate target appropriate for the individual. The week immediately following the full moon is a period of recovery, leading down toward the trough of low energy, and then the exercise cycling starts over again.

A person with a chronic disease would want to do the exercise cycles using a heart rate monitor to reach target heart rates, which rise steadily during the three "on" weeks, then rest during the fourth. A healthy individual might want to put up a chart at work showing the exercise pattern described in the previous paragraph. Then, guided by a lunar calender, he or she could take a break at the appropriate time to work out on the stairs or a trampoline, whatever is convenient, matching the intensity of the workout to the time of the month. You don't have to go to a gym and get togged up in exercise gear; as one of Dardik's colleagues put it, "This is the Jewish princess workout program because you don't have to break out in a sweat!"

"Compared with the way we used to do cycles, which was much more linear, not tracking circadian or lunar cycles, this new protocol had tremendous appeal to us," says Dardik. "Mapping the intensity with which we do the cycles onto the circadian energy wave, and mapping that onto the lunar energy wave, brings us as close to reconnecting to the rhythms of nature as we can imagine. And when you do it like that, you really do feel in harmony with nature's rhythms, which is what this program is all about."

People in the medical world began to hear of Dardik's exercise program and its beneficial impact on heart rate variability, partly by word of mouth and partly from the circulation of preprints of his *Frontiers* paper.

In February 1997, Michael Zanakis, director of research development at the Kessler Institute for Rehabilitation, based in New Jersey, contacted Dardik: Would Dardik like to give an informal seminar to some of his research scientists? Naturally, Dardik jumped at the chance, seeing it as a pathway for *his* rehabilitation into the mainstream. Zanakis wanted to explore the possibility of incorporating Dardik's exercise program into Kessler's own programs for rehabilitating patients with heart disease. Over the following couple of weeks, Dardik spoke at several such informal gatherings and quickly won supporters, most notably Stanley Reisman, head of the biomedical signal processing lab at the New Jersey Institute of Technology, who formerly had a joint appointment at Kessler.

Reisman, an engineer and mathematician by training, had spent two decades studying heart rate variability's source and association with health and disease, which made him one of the world's top experts on the subject. "Mike Zanakis invited me to that first gathering," remembers Reisman, "because he knew of my interest in heart rate variability, and that was what Irv was going to be talking about. I'd not heard about Irv before, but at the end of his talk I went up to him and said, 'You and I think a lot alike. I would like to work with you.'" That was the beginning of an extremely productive collaboration that continues to this day. "I couldn't put into words at the time of exactly what it was that so strongly resonated with me in what Irv was saying," Reisman says. "It was a powerful gut feeling that I couldn't ignore."

In early March, Zanakis was driving Dardik to the institute and said, "We're having a formal research seminar on the 12th; would you like to take part?" Dardik said he would. "I was sitting in the backseat, and I leaned forward and said to Mike, 'I want to make sure, Mike, that you know my history," recalls Dardik. 'I don't want any problems.' Mike interrupted me before I could finish and said, 'Don't worry about it. Everything will be fine.' I thought, 'OK, he must know about my license issue,' and left it at that."

Posters announcing the seminar were displayed around the institute, saying, "Heartwaves: The Solution to Heart Rate Variability." The poster described Dardik's background as vascular surgeon, winner of the American Medical Association's Hektoen Gold Medal, and founder of the

U.S. Olympics Sports Medicine Council. It ended by announcing, "The solution to HRV will be presented by relating the rhythmicity of human behavior to a new understanding of heart rate as Heartwaves."

"There I was, on the evening before the seminar, getting my material together, feeling good about what was to happen the next day, and the phone rang," remembers Dardik. "It was Mike Zanakis. He said to me, 'I'm sorry, Irv, but I just found out about your losing your license. I'm dis-inviting you from tomorrow's seminar because Kessler is a prestigious institution, and we can't afford a scandal. Good-bye.' I was shocked, because I thought he already knew from what I said to him in the car. I guess I must have assumed that he knew what I was talking about when I said I wanted to make sure that he knew about my history. Maybe he thought I was talking about the Olympics or something. I don't know."

Dardik immediately called Reisman and related what had happened. "Don't worry, Irv," said Reisman. "I don't care about the license. What you are doing is important, and I want to continue working with you, no matter what."

The day of the seminar turned out not to be quite what Zanakis had in mind. While he was introducing the seminar, a bevy of FBI agents swept in, handcuffed Zanakis, and took him to the courthouse. Zanakis was arrested for allegedly trying to extort $5 million from McDonalds by planting a fried rat tail in his son's fries at one of the fast-food chain's outlets on Long Island, claiming the food had been contaminated in the restaurant's kitchen. Earlier, he had persuaded Coca-Cola Company to pay him $4,600 to keep quiet about grease particles he alleged he found in a can of Coca-Cola Classic he drank.

This time Zanakis wasn't so lucky. The FBI analyzed the rat tail and found that it wasn't from a "town" rat but instead came from a strain of experimental rat that Zanakis used in his lab work. Zanakis was subsequently convicted of extortion and sentenced to 30 months in jail. So much for not wanting the name of Kessler to be besmirched by scandal!

Even though Dardik's hoped-for collaboration with the Kessler Institute came to naught, he did begin to work with Reisman. "My job," explains Reisman, "was to put into mathematical language what Irv intuited for coming up with target heart rates for individuals on the exercise protocol. We would look at someone's numbers, particularly the rate of acceleration and the rate of recovery of heart rate, and Irv would talk about his rationale for that person's target heart rates for the next session. I would come up with an algorithm that encapsulated mathematically what he was talking about. We went back and forth quite a lot with that until we came up with something that he and I were happy with."

The next step for Reisman was to build a Web site, so that a person could do cycles at home and download the heart rate data to the Web site. The data would be analyzed automatically, and the person could then retrieve the new target heart rate. That took about a year, into early 1998, and formed the kernel of what Dardik and Godfrey hoped would become a commercial business, delivering the exercise protocol to paying customers over the Internet.

"Irv," said Dick Fox over the phone, early in January 1998, "I have a friend, Sidney Kimmel, I think you could help. He had a mild heart attack a while back, he has high blood pressure, and he can't get his heart rate above 90. He's very depressed, no energy, just dragging himself about. What do you think?" Fox talked to Dardik for about 10 minutes, spelling out more of his friend's medical history and his present condition. Dardik said, "If he doesn't do something soon, he will be dead in six months. Have the guy call me."

Three days later, Kimmel called the farm and spoke to Dardik, who said the exercise program could help someone in his condition because it would increase heart rate variability. "I had no idea what he was talking about," Kimmel told me, "but that didn't matter because Dick Fox was so positive about Irv, and I really did need help. The regular docs were following the conventional approach to rehab, but it wasn't helping as much as they hoped. Irv told me where he lived, out in the boondocks, but I said I'd be there the next day."

The chauffeur-driven black Mercedes pulled up to the farm around noon, and out stepped Kimmel, with sleek gray hair and in a perfectly tailored dark suit and red tie, but with the appearance of a man whose life had been sucked out of him. "He looked like the walking dead," remembers Godfrey.

"Irv took me to his office in the old barn," recalls Kimmel. "It was a shambles, papers and books all over the place, and Irv looked to me like everything you imagine that an absentminded professor should be like: tousled hair, sloppy clothes, sneakers, and talking nonstop, going from one topic to the next without taking a breath. At the same time, I liked the guy; we clicked."

Dardik had Kimmel do some brief, gentle cycles to get some parameters. "His resting heart rate was 57, and his max heart rate was 87," recalls Dardik. "I could see that he was going down the tubes fast, with that kind of max heart rate. I said I would work with him."

At that point, Dardik had no idea who this new client was, beyond the fact that he was obviously successful in whatever business he was in. Only later did he discover that Kimmel was founder and chairman of Jones Apparel Group, a giant in the clothing industry. Kimmel's origins were humble, at best. "We were poor, and that's an understatement," he says. "By the time I was six years old we had moved about eight times, usually at three in the morning because we were in arrears with rent, and the only way we could survive was to move to another place and not pay the rent. Our family was obsessed with survival, and I was determined that when I grew up I wouldn't have a life of financial stress."

Kimmel was more interested in making money than in getting an education, and so after just a year at Temple University, he dropped out and looked for ways to make money. He went into the clothing business early on, eventually started his own company, and achieved his goal of financial independence, many times over. *Fortune* magazine recently reported his net worth to be around $1-3 billion, and he routinely makes it onto *Business Week*'s list of top 50 philanthropists. In 2004, he placed 20th. He gives to the performing arts, to Jewish foundations, and to cancer research, where his donations of more than $450 million around the country make him the single biggest donor in cancer research in Western civilization. "I feel I have an obligation to give back," he says simply.

A week after that first meeting at the farm, Dardik began making regular visits to Kimmel's office in Manhattan, working personally with him on the cycles program. "Irv took it slowly with me at first," says Kimmel, "but gradually, within four months, I was able to get my heart rate up to 120. My cardiologist said to me, 'Don't go past 120 while you are on beta blockers.' I told him that in that case I didn't want beta blockers anymore, and I eventually reached 150! I started to feel better. I started to look better. And there was more life in me." He kept up with the program, continued to make progress and, by the fall of 2004, didn't look like a man in his mid-70s and certainly not a man who had been on the brink of death. As his friend Dick Fox says, "Sidney looks great. It's an amazing story, his complete transformation."

Kimmel paid Dardik well for his work and put up money to establish the Dardik Institute, which he housed in a grand hillside mansion near Tewksbury, New Jersey, with magnificent southern views across Round Valley. Dardik, Godfrey, and their two children moved into the mansion in the late summer of 1999, and the idea was that it would be a center for visiting scientists to explore and expand the SuperWave theory. Very soon, however, Kimmel became more interested in the program as a business opportunity. LifeWaves International was founded with a business staff of half a dozen, with Godfrey as CEO and Dardik as chief scientific officer. The aim was to deliver the program to the public over the Internet, very much as had been envisioned following the work with Stan Reisman a year earlier. Dardik's and Godfrey's lives had taken a most unexpected turn. But Kimmel had an impact on their lives and on the validation of the protocol in another way, too.

In December 1998, Dardik decided it was time to get in touch with Ary Goldberger again and give him the "something concrete" that he said he needed in order to take Dardik's work seriously. "I talked to Ary on the phone and told him what had happened with the program since we last talked, in 1990," recalls Dardik. "I told him how the protocol had been changed substantially to be in sync with circadian and lunar cycles, and that we were getting funding for the work from a New York businessman. He responded very positively this time, and I said 'I have something I want to send you.' He said, 'OK, send it along. We'll do an analysis.'" That "something" was Kimmel's heart rate data, one set from the very start of his work on the program, and the second after nine months.

The aim of the analysis was to tease out the complexity displayed by the two data sets. It wasn't just heart rate variability, but the kind of mathematical evaluation that Goldberger had pioneered in his Laboratory for Nonlinear Dynamics, at Beth Israel in Boston. It was the type of analysis that is difficult if not impossible for a nonmathematician to follow, but the bottom line was simple enough to understand. "The first data set showed very low complexity, something you would get from an elderly person with chronic disease," Goldberger told me. "The second data set was dramatically different, with substantially enhanced complexity. It looked like someone who was 20 years younger than the first

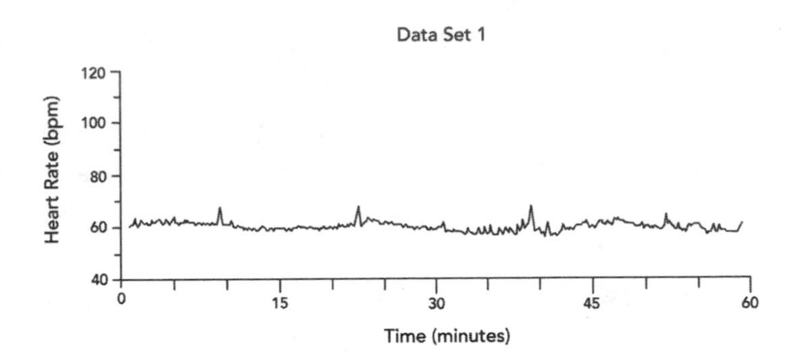

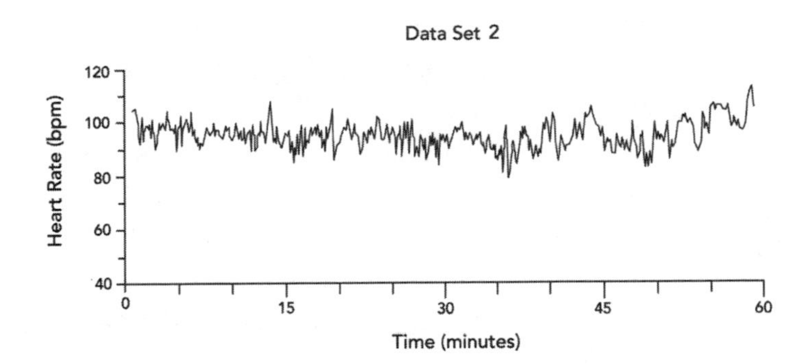

Effect of Cyclic Exercise on Complexity of heart function: Data set 1 shows an individual prior to the protocol, which shows low complexity associated with disease. Data set 2 shows the same individual after three months on the protocol, which shows increased complexity associated with health.

set, like a healthy middle-aged person. It looked like two completely different people, but it was from the same person, before and after the exercise protocol. I was impressed. This is not something cosmetic, like putting a Jaguar body onto a Hyundai engine, hoping you now have a car that can perform better. This is something fundamental, because when you see complex dynamics of the sort we had in the second data set, it is not that the heart just looks more youthful; the person *is* more youthful because these dynamics are coupled to the functionality of the system *at all levels*. These data from Kimmel were exactly what we needed before committing time to a clinical. Now we had real pilot data we could show to our colleagues."

Goldberger wrote to Dardik on February 4, 1999, giving the results and saying that he looked forward to the opportunity "to test and validate the efficacy of your innovative protocol in a controlled fashion." A year later, the trial began, and it included experienced researchers from Harvard Medical School, the New Jersey Institute of Technology, the State University of New York, and Columbia University in New York. The study was funded in part by the Centers for Disease Control and Prevention and by Kimmel. Eleven healthy nurses from Hunterdon Medical Center in New Jersey were the test subjects, and they embarked upon the cyclic exercise protocol for just eight weeks, using trampolines or stationary bicycles to do cycles three times a week, following the new protocol. Ideally, clinical trials have all kinds of controls built into them, to help ensure that the observed results are the unique outcome of the program being tested. In this case, the trial would have compared eleven healthy people on an eight-week program of conventional exercise with eleven healthy people on no exercise program at all, but under the same scrutiny for the same period of time. But, because of the extra time and extra cost that such controls would have demanded, they were not included in this preliminary trial.

"We had to make a decision about the scale of this initial clinical trial," explains Goldberger. "What we ended up saying was, look, this is a new type of exercise that most people don't know exists, and if they did they would probably be skeptical that it would work at all." The challenge of the trial, therefore, was this: "To determine whether a short cyclic exercise regimen would have *any* effect in certain measures of fitness and health in a population of healthy individuals." Fitness measures included people's ability to deliver oxygen to muscle tissue while exercising, or aerobic fitness. Health measures included heart dynamics, such as heart rate variability, immune defenses, and well-being, such as levels of anxiety, sleep patterns, and other quality-of-life issues.

The result, in the cautious parlance of science, as published in the March/April 2002 issue of the *American Journal of Medicine and Sports*: "The authors conclude that even very short (8-week) implementation of a cyclic exercise protocol may have beneficial effects. Further studies are indicated to compare cyclic and traditional protocols, both in healthy subjects and in selected patient groups." Translation: It worked much better than most people dared hope!

Fitness measures were up, with a 15.5 percent increase in the ability to metabolize oxygen; certain measures of heart rate variability increased 9 percent; blood pressure improved, with a 7.5 percent decrease in diastolic pressure; and immune defenses were boosted. The Mental Health inventory was "very positive," says Herbert Benson of Harvard's Mind/Body Medical Institute. "People had fewer symptoms of anxiety, and they felt better in themselves." A comment from exercise physiologist Rochelle Goldsmith about the impact on the efficiency of oxygen consumption can be applied to *all* measures in the trial: "It's amazing to see so much return *for so little exercise*." The increase in fitness that the nurses achieved was "what you would expect from people doing at least twenty minutes of exercise, three or four times a week, over a substantial period of time," says Goldsmith, head of the exercise physiology program at Columbia University College of Physicians and Surgeons, where the fitness tests were conducted.

The same is true for immune defenses, says George Stephano, an immunologist at the State University of New York, Old Westbury. "We know that moderate exercise can sometimes have beneficial effects on the immune system," says Stephano, "but these are often accompanied by indications of stress. In this clinical trial, you see positive effects on the immune system, but *no signs of stress*, and that has to be good." Another less scientific but equally striking indication of boosted immune defenses was the fact that during the period of the clinical trial, a very large proportion of the nurses at Hunterdon Hospital came down with a flu that was raging in the area. All the nurses on the exercise protocol stayed healthy.

As for feedback from the nurses themselves, most reported that they loved the program, although it was sometimes "a pain trying to fit the cycles into the day." All the nurses were healthy, though some were fitter than others. Although some of the nurses mentioned boosts in fitness, such as "Now I have no problem running up two, three flights of stairs to get to a patient with cardiac arrest, for instance," most of their comments concerned general well-being. "I used to get manic before my period," says Carol Fiorino, "but the cycles evened me right out." Stephanie Dougherty gained a greater ability to manage stress, to be able to do things with less distraction, more concentration. "Things are flowing for me now," she says. "Resolutions are happening with a kind of effortlessness." Donna Cole, who coordinated the program, was already a runner and therefore considered herself pretty fit. But even she says that "I feel better and have more energy throughout the day." Everyone reported sleeping better and feeling more rested. And oh, the dreams! Many of the nurses reported having vivid, positive dreams, the kind that, to paraphrase Shakespeare, "knit up the raveled sleeve of care."

These published results were also presented at several conferences on health. Reaction in the press was very positive. Meg Jordan, editor of *American Fitness* magazine, wrote an editorial titled "Time to Completely Rethink Rest and Rhythm," in which she said, "Amazing new research has been released that should flip the fitness world upside down." She quotes Goldberger as follows: "This study is exciting because it presents the first evidence that a novel exercise protocol designed to train both activation and recovery phases of exercise may increase cardiovascular fitness, heart rate variability and enhance mood in healthy subjects." Wire services, including the Associated Press, and many online sites picked up on Jordan's editorial, as well as other articles in which she enthused about the program, saying, "It's the way people were meant to exercise." Finally, the Aerobic and Fitness Association of America endorsed the novel exercise program.

The one negative note came in a book, *Ultimate Fitness*, by Gina Kolata, a science reporter for the *New York Times*. Kolata is a prolific writer who covers, among other things, medical topics. In this context, she takes a strong stance against all forms of alternative medicine, according to Sheryl Fragin, who wrote a critique of Kolata in the October 1998 issue of *Brill's Content*. "There's no doubt where Kolata stands on the subject," wrote Fragin. "All five of her stories on alternative medicine have been about hoaxes, and quack doctors, in stark contrast to health columnist Jane Brody, who recently urged readers to consider some of the same therapies Kolata ridicules."

It is therefore not surprising that when she first heard about the results of the healthy women study, Kolata writes in her book, her "first thought is that something has to be wrong." She goes on to say that from the beginning, she thought "the study that was being pitched to me was likely to be inaccurate." She made much of the New York medical board's revoking Dardik's license. But, as Ary Goldberger points out, "the idea here is not connected with what the New York board did. This is testing in a group of *healthy* people." In any case, Dardik was hardly flying solo here. There were eight highly reputable researchers besides him involved in the work.

Kolata cites two statisticians' concerns about the lack of a control in the study and their doubts that the results were statistically significant. "She obviously didn't understand the reason we did the study," observes Stan Reisman. "This was a pilot study, simply to see if we would get any effect at all with healthy people on a novel protocol over a short period of time. Pilot studies are often done the way we did ours, and she should have known that. The fact that we saw the changes in health and fitness measures under these circumstances encouraged us to do further studies so we can understand more completely what is generating the effects we see." As for the statisticians' comments that the results weren't statistically significant, Reisman states flatly, "That's nonsense. Our statistics are as rigorous as anyone's. The question is, where do we go from here?"

Sidney Kimmel had much the same question at the time: Where do we go from here? "I found myself in a quandary," says Kimmel. "Here I was, doing the cycles and benefiting enormously from them. At the same time, I am going around the country building cancer centers, so I have a lot of opportunity to talk to scientists and doctors. But anytime I would bring up Irv's SuperWave Principle and try to explain that it brings an entirely different perspective on health and disease, they would pooh-pooh the idea immediately, and the conversation wouldn't last four minutes. My gut tells me that Irv is right, but it was hard to hear all this negative stuff about it. I have to admit that I was constantly concerned that people would be talking about 'Kimmel's Folly,' referring to my support of Irv's thing. It became more and more clear to me that we needed something more to bolster the theory; better clinical trials, for instance, or perhaps something right outside of medicine. If we got that, then these people would have to listen."

PART THREE
Riding the Wave

10

Testing the Science, in Medicine

"The goal of life is living in agreement with nature."

—ZENO

WITH THE TURN OF THE MILLENNIUM, Dardik had for 15 years been pursuing the notion that waves are fundamental, not only to health and disease, but to the very nature of nature. As we have seen, that pursuit has been a journey of waves in its own way, a roller-coaster ride of hopes and expectations raised only to be dashed. Recognition that what he was doing was of profound importance would appear like a distant specter, only to vanish, repeatedly. Dardik is a patient man, and he had to be. He had no choice. But as the millennium dawned, even he was becoming frustrated, as was his biggest supporter, Sidney Kimmel.

Kimmel had gone out on a limb, as visionaries do, supporting and standing by Dardik and *his* vision when many around him saw nothing of merit and said so bluntly. Unless the 21st century brought clear validation of the SuperWave Principle and quickly, his fear of being the less-than-proud sponsor of Kimmel's Folly would be realized, and Dardik's support would vanish.

The early years of the new millennium, then, would be a search for validation not only in medicine but also in physics. When we say "validation," it is not in fact strictly accurate, in the sense of "proving to be true," because in science one cannot prove anything to be true. One can only prove something to be untrue. It is what philosopher of science Karl Popper called the act of falsification. Suppose, for example, you have a hypothesis that all sheep are white, and you are standing in front of a field of white sheep; this observation does not validate your hypothesis. The observation is consistent with the hypothesis, and that is all that can be said. If you see one dozen, two dozen, ten dozen fields of white sheep, then the likelihood of the hypothesis being true becomes stronger and stronger. But lurking behind a distant hedge might be a black sheep. See that one black sheep and you have proved the hypothesis to be untrue; it is falsified.

When, as we saw in the previous chapter, Dardik claimed that the literature on heart rate variability and on the harmful effects of sustained exercise *proves* that his theory is right about the health efficacy of cyclic exercise, that is not strictly true, either. Properly speaking, those observations can be said to be *consistent* with and strengthen the theory, and that is all you can ask for.

Dardik is a man of passion, commitment, and powerful intuition. That last word crops up frequently when people are asked to characterize him. It is his intuition speaking when he says "I know I am right" or "This proves I am right." Einstein said the following about intuition: "The intuitive mind is a gift, and the rational mind is a faithful servant. We have created a society that honors the servant and has forgotten the gift." Intuition is in fact the keenest tool in science, despite what the language of science purports.

Ironically, Kimmel apparently possesses the same kind of intuitive power in his own ventures. Against all odds, he defied his closest peers when they strongly advised him not to start his apparel business three decades ago with limited capital in a down period in the American economy. Today, Kimmel remains as founder and chairman Jones Apparel Group, a New York Stock Exchange company with annual sales approaching $5 billion.

In the realm of medicine, Dardik, encouraged by the healthy women's study, embarked on two further clinical trials, one with people with Parkinson's disease, the second with HIV-infected patients. The decision to conduct a clinical trial with Parkinson's patients didn't evolve from careful strategizing, but rather was born of circumstance, as so often happens in science. Dardik's interest in the origin of order and chaos led him to meet Curt Lindberg, who had established a nucleus of people interested in complexity science at VHA, Inc., a large health-care cooperative. His father, Bob, was suffering from Parkinson's disease, and when Curt heard about Dardik's cyclic exercise protocol, he was eager to see if it might alleviate his father's symptoms.

On July 5, 2000, Robert Lindberg Sr. attended the annual Connecticut Jazz Festival in the small town of Moodus, about two hours' drive from his home in White Plains, New York. "The festival has always been the highlight of the year for our family," says Bob Sr., who has a passion for music of many kinds, but especially jazz. At 78, his tall stature and strong facial bone structure betrayed his Scandinavian origins, but he was no longer as spry as he was as a young man. Accompanied by his wife, Lois, and one of his sons, Bob Jr., he used a cane to steady his balance during four hours of festivities. No great surprise there.

What was surprising was that at the previous year's event, he depended on a sturdy walker to get around, and then extremely awkwardly and for a short time only. "Although no one said anything, we all thought it was Dad's last time at the festival," Bob Jr. now says. "It was just remarkable."

Just a few months before the 2000 festival, Bob Sr. was in the grips of the usually unrelenting, progressive symptoms of advanced-stage Parkinson's disease, which had first been diagnosed back in 1988: namely, difficulty walking; the constant fear of "freezing" in the midst of trying to move; a low, mumbling speech; a wit, once sharp and humorous, now dulled and hidden behind a bland facial mask; a faded memory, frustratingly cheating Bob's attempts to play bridge with his friends or to do the daily *New York Times* crossword; a soaring singing voice, with which he rejoiced in life and Christ in his local church choir, gone flat, or "caterwauling," as he put it; and the torture of dark, ominous thoughts drifting through sleep-deprived nights.

"I knew I was going downhill fast," Bob told me when I met him in the fall of 2000. For Lois, it was painful watching the husband she loved slip away. "I thought our life together was over," she said, "our life of doing things together, going places together. It seemed over."

Then, in the April before the jazz festival, Bob embarked—initially reluctantly, it has to be said, because he feared it was beyond his physical resources—on Dardik's cyclic exercise program. This included running up stairs in a short burst (oddly, Parkinson's sufferers are often able to manage stairs easily, even though walking on a flat surface can be a challenge), a strenuous minute on a stationary bicycle, or walking quickly around the living room three times. By festival time, the program had not only restored the exercised muscles back to some semblance of normalcy, but also touched parts of Bob's being that seemed distant from the program itself. His singing voice returned; his wit, his smile, his facial expression reemerged; he no longer had problems with bladder control; and he was able to sleep well for the first time in years, no longer struggling with the dark thoughts of night.

Here was a man who had been, as he put it, "accommodating my schedule to my health," now working in his yard again, more or less when he wanted; making trips with his wife again; going to church again, and singing. His bridge friends no longer had to pretend that he hadn't forgotten trumps, because he no longer did. This wasn't a miracle cure; Bob still suffered symptoms of Parkinson's, but the disease no longer seemed the "sentence to purgatory" that it once had been. "Even if Bob doesn't improve anymore, it's been such a wonderful experience," said Lois.

Curt thought about his father's improvement in terms of time. "Dad was like he was three years before he started the cycles," he said. "And that made a huge difference. It bought him more time with my mother." Bob Sr. died in September 2001, of complications from advanced prostate cancer.

Marilyn Rymer, a neurologist at St. Luke's Medical Center in Kansas City, another in Curt Lindberg's complexity circle, heard about his father's story and was impressed, but also cautious. "What I saw was a pretty interesting and dramatic improvement in a patient who was chronically debilitated," she recalls. "The question for me was, Is this the result of a patient getting better simply because someone was paying attention to him, the Hawthorne Effect? Or does the protocol have real medical benefits? I thought that should be tested."

Dardik and Godfrey had already started to plan a clinical trial at a New Jersey hospital at which they had some contacts through the healthy women's study. But the hospital was in the midst of a merger, and the subsequent bureaucratic snarl put an end to that venture. So when Rymer offered to step in and run a trial in Kansas City, they were delighted. Curt Nonomaque, president and CEO of VHA, was enthusiastic about the project and arranged a half-million-dollar grant to support it. The Kimmel Foundation provided much of the rest of the necessary monies. Clinical trials, even small ones like this one was to be, are very expensive.

"We began with 19 people, 15 men and 4 women," says Rymer. "The trial ran for 12 weeks initially, beginning in May of 2001; and then another 12 weeks, beginning in October. We followed the protocol of three weeks doing cycles, one week rest, and so on. They exercised on a stationary bike, a trampoline, whatever was appropriate for the individual. Several people were in wheelchairs, so we had to start slowly. Many of the people came from many miles away, so they did a lot of traveling each week. The remarkable thing was that only one person dropped out, and that was made even more remarkable because it was the coldest November and December on record for Kansas City!"

The results were uniformly positive. On the biochemical level, there was an increase in blood levels of anti-inflammatory molecules, such as interleukin-10 and adrenocorticotropin, which indicates a boosted immune system and a healthier body. This observation, which was also seen in the healthy women's trial, is especially significant, not just in Parkinson's patients, but in patients with any kind of chronic disease. Immunologists now know that inflammation is a common feature of chronic diseases and contributes to their degenerative aspects.

Here's the key point: Sustained exercise is known to *increase* pro-inflammatory molecules in the blood because the body is under stress.

That's a bad thing to induce in a healthy body, but it is especially harmful when disease, even incipient disease, is present. On the other hand, the cyclic exercise protocol *reduces* the inflammatory response by boosting anti-inflammatory molecules. In a paper in the *International Journal of Molecular Medicine*, Rymer and her colleagues wrote that this effect "is associated with alleviation of some of the clinical characteristics of Parkinson's disease." It is therefore easy to see why, at this molecular level, cyclic exercise would be beneficial in *any* chronic disease, as well as why sustained exercise would be harmful.

Rymer and her colleagues also saw improvements in blood pressure, in the range between resting and maximum heart rate, and in heart rate variability, just as there had been in the healthy women's study. "These results were very encouraging," says Rymer, "but the changes in people as people was much more dramatic. We had people who were able to walk unassisted for the first time in years. People who had had the typical poker face you get with Parkinson's were smiling and talking normally for the first time in years. It was amazing."

The researchers captured some of their patients' experiences on tape:

Patient A: "I can see a lot of difference with this study. I used to take a cane with me everywhere I went. Some things I just couldn't do, and some things I didn't have the energy to do. Since these exercises I do pretty much what I want to do, or at least what the wife tells me to do. [laughs] This last year I've done woodwork, concrete work, kept my yard cleaned up and mowed, and done things around the house that I wouldn't have tried to do a year ago.

"Last week I took out a doorway in a wall, and made shelves. Moved the door over is what I did. I can do things now that I couldn't even try to do. I work on the car—brake work, light maintenance, change the oil, things like that."

Voice-over: "You'd given up those things?"

"Pretty much. I would have my grandson-in-law do things, like changing a headlight. Seems simple, but a year ago I wouldn't have the energy or the confidence in myself that I could do it. This is where it has helped a lot, confidence."

Patient A's wife: "Before he did the program, we were planning to widen the doorway so we could have a wheelchair go through it. We scrapped that idea. We were going to have a ramp; there were three steps to get up into the house. We were trying to figure out a ramp, so we could get a wheelchair in there. We scrapped that, too. It's not even an option right now."

Patient B: "I know that doing the cycles, I have gotten my energy back, and I feel better. The zest for life that I didn't have when I started the cycles, it's back." Patient B's wife: "When we entered this program, Chuck was almost devoid of any facial expressions, barely moved, would sit in his chair all day and all evening, not motivated by anything, which was totally unlike his personality. This went on for quite a while, and I began to see I was losing my husband. He was becoming more and more withdrawn, more silent, and all the expression from his face disappeared, and he lost his wonderful sense of dry humor. When he started the program, I slowly began to see my husband come back to life. One day I walked into the family room, and he was smiling. I couldn't believe it. I thought, 'He's come back. Wherever he's been, he's come back.' And he's been doing that ever since."

Voice-over: "We had asked this man if there was anything he hoped to be able to do now that previously he could not. He said, 'I want to dance with my wife.' After eight weeks, he did that!"

"These were some of the better results," explains Rymer, "but they were typical of the direction of progress in the group as a whole. This wasn't a controlled study, with people doing regular exercise, for instance, so we don't have a direct comparison. But from my experience and my reading of the literature, I would venture that cyclic exercise has more efficacy than regular exercise, probably significantly more." Without proper control, it is difficult to be certain, of course, but the results are *consistent* with the predictions of the SuperWave theory.

In July 2001, Dardik read an article on the Salon.com Web site about the work of Luis Montaner, head of HIV research at Philadelphia's Wistar Institute. The article described Montaner's recent studies, which showed that when HIV-infected patients took their drugs intermittently, as opposed to continuously, their immune defenses were significantly boosted. "I called Montaner out of the blue," recalls Dardik, "and said, 'I am doing something that I think explains what you are seeing. It has to do with the beneficial effects of cycles, and what you are doing is cycling the drug therapy.'"

That piqued Montaner's interest. "The idea of cycling back and forth is very much in the mind-set of people thinking about the immune system," Montaner told me. "So when Irv talked about the concept of cycles as being a primary basis for regulation in the healthy body, and that the results of intermittent therapy might be another expression of that, I was very receptive. I said to Irv, 'That's worth pursuing. Why don't you come to Wistar so we can talk some more about it.' We e-mailed back and forth a bit, and Irv filled me in on the other work he had been doing, saying that the exercise protocol is the same, no matter what disease you are interested in. We agreed to meet on the fifth of September."

The meeting was very productive, and it set in motion plans for a clinical trial that would be conducted in conjunction with Philadelphia FIGHT, the Field Initiating Group for HIV Trials. For reasons having to do with federal regulations current at the time, the patients had to continue taking drugs throughout the trial, which limited the ability to test the impact of the exercise protocol on the level of the HIV virus and white cell counts in patients' blood. "We decided to focus primarily on quality-of-life measures," explains Montaner, "and that's not trivial, because anti-HIV drugs are quite toxic, and patients suffer a lot of detrimental effects, such as lack of energy, and depression."

The trial began the following April and ran initially for eight weeks. For the first time in trials of the efficacy of cyclic exercise, there would be controls. FIGHT helped recruit 52 patients, who were assigned randomly among three groups: the first did the cyclic exercise protocol; the second spent an equal amount of time doing traditional aerobic exercise; and the third did no exercise. "The results were quite impressive," says Montaner.

"There's no doubt that those people in the cyclic exercise group did better than people in the two control groups."

For instance, when asked to rate "how much I enjoy life," people in the cyclic exercise group moved from "some of the time" at the beginning of the trial to "most of the time" at the end of eight weeks. Responses to the statement "My sleep was restless" moved from "most/some" to "some/rarely." Similarly, these participants reported that it took much less effort to do daily activities by the end of the trial, they felt their lives much less of a failure, and they felt generally much happier. "About a month into the trial, FIGHT organized a public meeting to describe what they were doing for the community," says Dardik. "I was asked to be there. At one point, a man from the audience went up to the microphone and explained that he was on this clinical trial. He was in tears, and he said, 'For the first time in so long, I have been able to go to the park and walk with my kids, because before I didn't have the strength to do it. Now I can!' It was very moving, I can tell you."

Jane Shull, executive director of Philadelphia FIGHT, was distinctly enthusiastic about the program. "Quality of life is a very important issue in this community," she said at a June 2003 meeting announcing the results. "Up until now, we have only been able to offer medication to suppress the virus, but the side effects take a toll on the quality of life. If patients can reduce the side effects of their medication, we have overcome a major obstacle. And it looks as if this cyclic exercise program can help patients do just that." Ever the cautious experimentalist, Montaner told me, "The quality of life improvements are significant and real, but we don't know whether it is a result of a biological effect of the cycles, in the way Irv's theory would predict, or whether it was a psychological effect, a placebo effect. But the results are interesting enough that I should like to find out." Again, we can say that the results are *consistent* with predictions of the SuperWave theory.

Recall how Tony Schwartz ended his article about Dardik's work in *New York* magazine: "Dardik's unified theory isn't likely to win wide acceptance by the scientific Establishment any time soon. What's harder to dismiss is his clinical success in treating patients with chronic diseases. If he can demonstrate that his program is as effective for patients in controlled studies as it has been anecdotally, even skeptics will feel compelled to take a closer look at his broader ideas."

The two studies presented here, limited though they are, do bolster Dardik's argument about the efficacy of making waves and their impact on health.

11

Testing the Science, in Physics

"We shall affirm that the Cosmos, more than aught else, resembles most closely that Living Creature of which all other living creatures, severally and generically, are portions. For that Living Creature embraces and contains within itself all the intelligible Living Creatures, just as this Universe contains us and all the other visible living creatures that have been fashioned."

—PLATO

THE UNIVERSE ITSELF IS, of course, the biggest example of wave motion in nature, gigantic in size, unending in time. Some 14 billion years ago, an incomprehensibly dense ball of energy and matter exploded, an event known as the Big Bang. That event set in motion cosmic processes that spawned stars, planets, and galaxies that formed our present-day universe as it expanded into the void that is space. No one knows when, but this expansion will one day likely end, and the reverse process, collapse, will set in, bringing all of matter to a single point once again. The stage will then be set for another expansion, and another collapse, repeatedly through an infinity of time—a repeated rhythm of life and death on the grandest scale of all.

One of the most remarkable findings in cosmology of the second half of the 20th century was the discovery in 1992 of the source of structure in the universe, the origin of galaxies as they distilled from the debris of the Big Bang. An echo of that fireball is known as the microwave background radiation, a dim afterglow of energy spread throughout the space.

But there was a puzzle. If energy and matter were spread evenly throughout the universe, as can easily be imagined in the aftermath of a giant explosion, how did matter clump together to form galaxies? The 1992 discovery, by George Smoot of the University of California, Berkeley, was that this afterglow is *not* smooth and continuous but instead is regularly wavelike. This waviness was apparently the template upon which structure was constructed, with matter coming together at the peaks of the waves, as embryonic galaxies. Smoot described the waves as "wrinkles in time" and said it was "like looking at the handwriting of God." A little hyperbolic, perhaps, but a significant enough finding for Stephen Hawking to describe it as "the scientific discovery of the century, if not all time."

Planet Earth sits in an obscure part of our galaxy, the Milky Way, which, by the most recent estimate, formed some 13 billion years ago, a majestic pinwheel of stars and gas. Until very recently, cosmologists spoke of our galaxy's formation in the past tense, as if it were a fait accompli, finished business. But as cosmologists Brian Walker and Phillip Richter wrote in the January 2004 issue of *Scientific American*, it is anything but. In their article, titled "Our Growing, Breathing Galaxy," they describe the galaxy, and all galaxies, as still in the process of forming. The Milky Way contains about 100 billion stars, distributed in a disk some 100,000 light-years across and 3,000 light-years thick—and growing, fueled by gas and remnant galaxies it swallows up. "The galaxy is breathing," they write, "pushing out gas and then pulling it back in, as if it were exhaling and inhaling." Rhythmic, wavelike, as if it were a living organism.

The history and future of the universe are written in the language of waves, from its inception to the formation of galactic structures through space, to the life of individual galaxies themselves. Waves are fundamental to the very being of our universe.

Dardik's attempts to find a sympathetic ear among physicists for his theory had been mixed, at best. Even among those who did evince an interest and support, nothing tangible occurred, no recognition that the SuperWave Principle was new and important, no research program to test it. With the Bar Mitzvah of Trevor, Dardik's and Godfrey's son, that would begin to change—in an utterly unpredictable way.

Midway through 1998, Godfrey engaged a young rabbi, Samuel Suffran, or "Shmuli," as he was known, to prepare Trevor for his important rite of passage. Shmuli became close with the family and spent a lot of time at the farm, talking about the history of Judaism and the meaning of upcoming festivals. The whole family was involved.

"A couple of years into the relationship, Irv started to tell me his ideas about waves, and it immediately clicked with me," Shmuli now says. "It wasn't just the science, because I'm not a scientist, so I couldn't really evaluate that. It was that the whole idea was so like Judaism itself, where waves and cycles of repetition are so important. A fundamental part of Judaism is reconnecting the spiritual with the physical, like Irv's focus on health and the body, and Judaism's with celebrating cycles. I said to Irv, 'I know of a physicist who might be interested in what you are saying. Would you like me to try to make contact with him?' Naturally, Irv was enthusiastic about the idea, but it would take a while before anything would happen."

The physicist Shmuli had in mind was Herman Branover, a Russian-born Jew, an expert in turbulence and a professor at Ben Gurion University in Beersheba, Israel. Although he was raised in the spirit of atheism in Russia, he became very religious in the ultraorthodox Lubavitch sect and eventually devoted a lot of time and energy to bringing Judaism back to Russia. This included translating the Jewish holy scriptures into Russian, a project funded by Sidney Kimmel. A man of considerable stature in the Jewish community, Branover was in great demand to speak at synagogues wherever he went. One of his frequent stops was in London, where a close friend, Mordechai Suffran, Shmuli's father, was a prominent rabbi. Shmuli, who came to the United States in 1992 to go to rabbinical school in Morristown, New Jersey, met Branover on a number of occasions and so knew of his scientific interests.

"Because of his reputation in science and religion, Branover is always being approached by people who want to talk to him," says Shmuli, "and my friends said, 'He's been introduced to too many people; don't bother him.' But when I talked to my father and explained what Irv was doing, and how it was so in tune with Judaism, he said, 'Call Herman. Let him decide if he wants to see Dr. Dardik.' So I called Branover, explained that

I had been working with Irv's family, and told him about the waves idea. He quickly became very excited, and we had a long discussion about the similarity between Irv's concept and Judaism. In the end he said, 'Tell Dr. Dardik to call me sometime.' It wasn't just the science that Branover responded to. It was a spiritual connection, too."

"Early in February of 2001, Irv phoned me from the States and gave me an hour-and-a-half lecture on his theory," recalls Branover. "I was very impressed with what he said, so I suggested that we should meet as soon as was possible. I am chairman of the organizing committee of the biennial world conference on religion and science. The more I study both, the more I find harmony. The next meeting of the committee was to be the following month, in Florida, so I suggested to Irv that he should try to come down on the 13th. He said he would."

"Alison and I were scheduled to be working with Sidney Kimmel in Palm Beach, Florida, at the time Branover suggested we meet, so it seemed like divine intervention," recalls Dardik. "We were staying at the Four Seasons Hotel, so we suggested we meet there. We tried to get a conference room, but they were all taken, so we ended up having our meeting in a cabana, by the pool. Branover was very much the part of a Labovitch Jew—the coat, the beret, long white beard. He's in his 70s and slightly stooped. What a crew we were, kids running around, jumping in and out of the pool.

"Anyway, we organized a flip chart, and so there I was, delivering a long spiel about SuperWaves in this open-air cabana. I was thinking to myself, 'Does he know what I'm talking about? Is he really listening?' Every now and then he would say, 'That's interesting,' and I thought, 'Oh no, Niels Bohr used that expression whenever he meant, 'I could care less.' But Branover asked questions from time to time, so I just continued talking, for four or five hours!"

"Irv was talking about cycles and wave processes dominating nature, particularly in the human body," recalls Branover. "He talked about a hierarchy of waves, five, six, maybe seven levels. Then he suddenly switched to nonorganic processes, heat and mass transfer, conversion of energy, quantum mechanics, saying it is all dominated by waves. It was hard for me to agree with him immediately because he touched on so many complicated things. He did sound a little crazy at first, but, you know, you have to be a little bit crazy to develop something new and worthwhile.

"I kept asking questions, and he gave me good answers, even though they sometimes were a little bizarre. But someone once said, 'Whoever is not astounded by the strangeness of quantum mechanics doesn't understand it.' So I was not put off by the strangeness of what Irv was saying. And a lot of what he was saying about waves seemed natural to me because of my experience in turbulence. At one point, he talked about low-energy nuclear reactions, or cold fusion, as an example of where he thought waves were important. I had never worked with cold fusion, but I knew the story, of course. At the end of that meeting, I was convinced that Irv's thinking was new and important, and we agreed to set up a program to test the waves idea in cold fusion experiments." Dardik had not planned that outcome. The topic had simply come up as an example of waves in physical systems, but it had seemed so compelling to both Dardik and Branover that the notion for a project to test the idea emerged fully formed.

Branover was so energized by the conversation that he immediately phoned Ehud Greenspan, a professor of physics at the University of California, Berkeley. Greenspan, a world expert in hot fusion and nuclear energy, had been a colleague of Branover's for many years at Ben Gurion. "I told Ehud that Irv wants me to set up a team to explore the use of SuperWaves in cold fusion. I asked if he would like to help me with it," Branover recalls.

Greenspan's response was immediate and unequivocal. "Don't get involved with this," he said. "It is dangerous because of the history. You are older than I am, Herman, and you can do anything you like. But I'm younger, and I don't want to ruin my scientific career."

Branover persisted. "Think about it, Ehud," he urged his friend. Greenspan reluctantly agreed that he would, mostly out of respect for his former mentor.

At this point in the story, we should step back and fill in some background. Fossil fuel—coal, oil, and natural gas—upon which the United States depends for about 85 percent of its energy, is a finite resource, destined one day to be depleted. Nuclear reactors, which derive heat from the fission, or splitting, of heavy elements, are deemed by many to be unsafe for large-scale energy generation, and the by-products are dangerous and difficult to store safely. What of the opposite of fission—fusion of light elements—as an unending source of energy? This is what fuels our sun, where hydrogen atoms, at incomprehensibly high temperatures, fuse, yielding helium and yet more energy. Physicists in the United States and Europe have for decades been wrestling with trying to tame this process, known as hot fusion, in the laboratory. Stout vessels are filled with deuterium, an isotope of hydrogen, which is heated to 300 million degrees Celsius (more than 10 times hotter than the center of the sun). Under these conditions, atomic nuclei overcome the powerful force that normally keeps them from flying apart from one another, and they fuse, to release energy. The best results so far have been a release of about 60 percent of the energy consumed in the process, hardly a formula for delivering infinite energy.

Physicists have been pursuing this dream for more than four decades, all along promising that the longed-for infinite energy is just a few decades distant, but with no real success in sight. The hot fusion community is well established in science and ensconced in the political bureaucracy. To date, for instance, the U.S. Department of Energy has shelled out some $20 billion on this distant dream, with current funding running at around $500 million a year. When, on March 23, 1989, two chemists at the University of Utah announced that they had tamed the fusion monster, but in a small glass jar on the top of a lab bench and at room temperature, hot fusion protagonists suddenly saw an unwelcome rival for DOE funds. When a suggestion was made by a congressional committee that $25 million—a tiny fraction, compared with hot fusion support—might be set aside to explore the newly announced discovery, hot fusioneers circled the wagons to make sure it wouldn't happen. They succeeded.

The two chemists were Martin Fleischmann, a highly respected electrochemist from England, and Stanley Pons, chairman of the chemistry department at the University of Utah. The two had labored in secret for five years, spending $100,000 of their own money, on what was often described as a "simple" experiment that "any kid could do in high school chemistry class." It is true that the so-called electrolytic cell that

Fleischmann and Pons used in their experiments was of a type with which chemists had been familiar for decades. It consists of a glass cylinder, filled with heavy water (water in which most of the hydrogen atoms have been replaced with deuterium). There are two electrodes in the cell, a platinum anode at the edge, and a palladium cathode in the center. When a current is run between the two electrodes, the water is "split" into oxygen ions, which go to the anode, and deuterium ions, which migrate to the cathode. Some of the gases bubble up to the surface and are vented away.

The whole cell is encased in a calorimeter, which measures the temperature of the cell to a high degree of accuracy. Fleischmann and Pons observed that when the cell was running, a significant amount of heat (called excess heat) was produced, at least 10 times as much as could be accounted for by normal chemical reactions. They concluded that some of the deuterium penetrated the palladium cathode, where fusion occurred, releasing heat—a tiny-scale version of what happens in the sun. On that March day of 1989, they called a press conference to announce their results, which provoked a media storm of gigantic proportions. They were hailed in the press as potential saviors of humankind, delivering as they had the possibility of endless, safe energy. The University of Utah immediately announced a $5 million grant to establish a cold fusion research center.

The event also provoked a scientific storm of equal proportions, at first quite positive, even adulatory. Scientists across the western Hemisphere rushed to try to replicate the remarkable observation made in Utah. Very soon, scientists at laboratories in prestigious institutions such as the Massachusetts Institute of Technology and the California Institute of Technology, as well as Harwell in the United Kingdom, reported failure to replicate. Adulation quickly gave way to excoriation, and Fleischmann and Pons were denounced in the most unforgiving terms, to put it kindly. At a meeting of the American Physical Society held in Baltimore in May 1989, Stephen Koonin, a theoretical physicist from Caltech, had this to say: "It is all very well to theorize how fusion might take place in a palladium cathode. . . . One could also theorize how pigs might fly if they had wings, but pigs don't have wings." Others at the meeting described the Utah chemists as "hucksters" and "incompetent and possibly delusional."

By September, Fleischmann and Pons had virtually no support among their peers; by November, the case was closed when a Department of Energy panel issued a report that essentially said that cold fusion was an illusion. One of the panel's co-chairmen, John Huizenga, went on to write a book titled *Cold Fusion: The Scientific Fiasco of the Century*.

There are many facets, plots, and subplots to this story, chronicled in more than a dozen books. Some of them followed the path blazed by Huizenga, while others were positive, such as Charles Beaudette's *Excess Heat: Why Cold Fusion Research Prevailed*. One of the problems upon which many would-be replicators stumbled was the fact that the electrolytic cell that Fleischmann and Pons developed was far more tricky to work with than they were prepared to entertain. Critics in the hot fusion community contended that the experiment was obviously false because the nuclear products emanating from the system were not those typically seen in hot fusion reactions. They refused to consider that the reactions in cold fusion might be different and therefore produce different by-products. As a result of all this, government funding for cold fusion research vanished in the United States; prestigious scientific journals, such as *Science* and *Nature*, refused to publish papers on the topic; and the U.S. Patent Office declined even to consider patent applications.

Despite all this negativism among establishment science, individual researchers in the United States, Europe, and Japan persevered, often under the radar screen, particularly in the States. Many instances of excess heat production (that is, more energy comes out of the reaction than is used to drive it) were reported within the community, though not so reliably nor dramatically. But the fact that excess heat appeared to be a reality that demanded explanation encouraged cold fusioneers to continue their efforts. By the turn of the millennium, more than 600 papers had been published on the topic—in "lesser" journals, of course. In the fall of 2004, the U.S. Department of Energy, urged by some respected voices in the cold fusion community, such as Peter Hagelstein of MIT and Mike McKubre of SRI International (formerly known as Stanford Research Institute), agreed to reassess the situation. Although its conclusion fell short of enthusiastic endorsement of cold fusion science, it at least opened the door to possible government funding. Cold fusion was no longer the scientific pariah that it had been for close to two decades.

Thus, Greenspan's initial reticence to venturing into the cold fusion arena was quite understandable. Like many people, he had simply stopped paying attention after Fleischmann's and Pons's so-called pathological science was roundly slammed so soon after it burst onto the scene. He assumed that the science was dead and buried. Prompted by Branover, Greenspan began combing through the literature on cold fusion and talking to some of the people active in the research. This included Mike McKubre in Menlo Park, just an hour's drive south of Berkeley. "I was very surprised by what I read and heard," Greenspan recalls. "It seemed to be much more sound than I had assumed. I became sufficiently interested that I called Herman and agreed to be part of his project."

Dardik, meanwhile, had been talking with Sidney Kimmel about the proposed project with Branover and the enormous commercial potential of developing a cheap, virtually endless source of energy. Kimmel was immediately hooked and told Dardik to arrange a meeting with Branover and Greenspan. The meeting was set for June 8 in Kimmel's Manhattan office on 40th Street and Sixth Avenue. Greenspan flew in from Berkeley the day before and met with Dardik in the lobby of the Marriott, not far from Kimmel's office. "Irv talked at great length about his theory and how it applied to cold fusion," says Greenspan. "But for me the bottom line was simple. I said to Irv, 'If there is a 1 percent chance that this might be right, it is worth pursuing.' That night I could barely sleep at all, going over all the issues involved, what needed to be done, who might be involved. The more I thought about it, the more promising it seemed to me." The meeting the following day included Kimmel, Branover, Greenspan, Dardik, and Godfrey. Branover described the proposed scientific program and explained that although he and his people had not previously worked in cold fusion, they were very familiar with the technology involved, such as how to set up electrolytic cells. Kimmel had great respect for Branover because of both his scientific credentials and religious convictions and was obviously feeling very positive about the project. But it was something Greenspan said that clinched it for him. Greenspan talked about his restless night and said, "I initially thought this thing would be worth doing if there were a 1 percent chance that it might be right. But now, after thinking about it all night, I think there's a 50-50 chance it could be right." Kimmel immediately proposed establishing a company, to be called Energetics Technologies, to explore the idea of applying SuperWaves to cold fusion, and he asked Branover to get a team together to set up a lab in Israel. Everyone agreed that the project should remain under wraps, at least for a while, because of the magnitude of the commercial potential involved.

As soon as he was out of the meeting, Branover called Shaul Lesin at Ben Gurion. Lesin, a mechanical engineer who was born in Argentina and moved to Israel with his parents when he was 11, joined Branover as a technician in 1978. While in Branover's lab, he continued his studies as a mechanical engineer and obtained a PhD in liquid metal magnetohydrodynamics, a method for converting heat into electricity. Recalls Lesin: "Herman called me late in the evening and said, 'Shaul, I've met a medical doctor, Irving Dardik, who has a very interesting idea about something I can't talk to you about over the phone. And I met an investor who will back the idea, but I can't tell you who it is until we meet in person. They are planning to establish a lab in Israel to develop this new work. Are you interested in being part of it?' It seemed all very cloak-and-dagger, something I wasn't used to. I said to Herman I might be interested, but that we had to talk when he got back to Israel."

Like Greenspan, Lesin had paid no attention to cold fusion for years, and he wasn't particularly put off when Branover told him that some people likened it to alchemy. "I took it as a technical challenge to see how we might be able to make it work," says Lesin. "I thought about potential employees and prepared a budget for the proposed lab. It came to about a million dollars, for rent, salaries, and instrumentation." Everything moved very quickly. Branover decided to locate the lab at

Omer Industrial Park, near Beersheba, partly because his home is close by, but also because the park specializes in entrepreneurial high-tech ventures, which matches the spirit of Energetics. The park rests on a rise, surrounded by the Negev desert, 30 miles distant from Masada, King Herod's 2,000-year-old Northern Palace built atop a huge mesa, his last holdout. Lesin is a master organizer, and he had to be. "We started from scratch, nothing, zero," he recalls. "No scientific knowledge in the area, no lab, no team. By the end of 2001, the lab was ready and the beginnings of a scientific team were in place. We invited Giovanna Selvaggi, from the Low Energy Nuclear Reactions lab at the University of Illinois, to come for a couple of months and help us set up the first electrolytic cells. We were ready to roll."

Up to this point, the Omer lab was simply going over territory traversed by labs all over the world. What made this approach unique was the manner in which the current would be delivered to the cell. In every other lab, the current is a simple direct current, a steady voltage, or perhaps a simple sine wave. At Omer, the current would be modulated, in the form of a SuperWave, with three nested waves: a long master wave; superimposed on it, five smaller waves; and on top of those, five more yet smaller waves. Their hypothesis was that if SuperWaves are fundamental to nature, then inputting the current in a SuperWave form should enhance the loading of the deuterium onto the palladium and boost the fusion process.

Every Sunday morning the team got together to brainstorm, rehashing the previous week and planning the experiments for the next. One of the first decisions was to change the design of the electrolytic cell because nothing was happening, in terms of producing excess heat. By the end of 2002, the cell was producing excess heat, but not enough to get excited about.

At the same time, a second experiment was set up; namely, glow discharge cells, which other researchers had tried in other countries with limited success. Arik El-Boher, a longtime colleague of Lesin's at Ben Gurion, was recruited to take charge of this project. The cell is a stainless steel cylinder about six inches in length and a couple of inches in diameter. Its inner surface is coated with palladium, and a tungsten wire runs down the center. The entire apparatus is immersed in a water-filled calorimeter, which measures the heat generated by the experiment. The cylinder is filled with deuterium gas, and a current is then passed through the wire, which ionizes the gas and induces the deuterium molecules to migrate to the palladium walls. The deuterium penetrates into the palladium where, with luck, fusion takes place.

"We were ready to test the glow discharge apparatus early in 2003," says Lesin. "It was a Thursday afternoon, and we started up the cell for the first time, with a simple current input. We immediately started to generate heat, not much, but enough to see that the cell was working as we had hoped. We then switched the current to the SuperWave form, and right away the heat output jumped to 3.7 times the energy we had put in, which is far more than anyone had ever achieved with the glow discharge cell. By now it was nine in the evening, just Arik and myself in the lab. We were ecstatic, of course, and we were literally dancing in the street. But we were worried because we don't work on Saturdays, and we didn't want to leave the cell unattended over the weekend, so we shut it down.

"On Sunday morning we started the cell up again, but it was cold, nothing happening. We took the cell apart and found that the palladium had sloughed off the cylinder walls. We tried all kinds of things with the palladium, and we did get excess heat a few times, maybe 80 percent, 100 percent, but we were not able to repeat that first big output, not for a while anyway."

Lesin and his colleagues weren't particularly worried. They were confident they would eventually overcome the repeatability issue that bugs so many cold fusion experiments. In any case, they had a result they could crow about, and they began to discuss whether they should present it at the upcoming International Conference on Cold Fusion, to be held at the Massachusetts Institute of Technology in Cambridge that August. They were eager to stake their claim of an extraordinary result. But the downside was that patents were still pending on the use of the SuperWave in the system, and they were nervous lest others might steal the idea before the patents were secured. (Dardik had actually applied for a patent using waves in cold fusion back in the early '90s, shortly after Fleischmann and

Pons announced their results. It was denied "because there is no evidence that cold fusion is a reality, [therefore] no patents will be granted." The new patents were an update of the original application.)

The cold fusion community holds international conferences every 18 months or so, but the MIT conference was to be especially important. This would be the first time since the conferences began in 1989 that it was held in association with a prestigious U.S. university, such had been the pariah status of the science in this country. Its chairman, Peter Hagelstein, an MIT professor in electrical engineering and computer science, was a big name not only in cold fusion, but also in the traditional scientific establishment. He had been a protégé of Edward Teller, the father of the hydrogen bomb; he is the youngest recipient of the Lawrence Award for National Defense, bestowed for his invention of the X-ray laser, which was the core of Reagan's Strategic Defense Initiative; and he was part of the U.S. Department of Energy's community.

In the May before the conference, Branover and Dardik had still not decided whether they should present their glow discharge work, but they wanted to talk to Hagelstein about it anyway. Branover called Hagelstein, and a meeting was arranged for May 11. Dardik presented his SuperWave Principle to Hagelstein, whose reaction was mixed. Hagelstein is a theorist as well as an experimentalist, and for him a theory in physics is not a theory if it is not bolstered by a mathematical formulation. Dardik's theory is mathematics free. "Irv talked about rhythms in nature, and in health and disease," Hagelstein told me. "I am a big fan of what he says in that respect. And when he talks about having a systems-level idea to apply to dynamical systems, I'm OK with that, too. But given that Irv doesn't have equations, models, or experiments, then one can't take what he is saying overly seriously."

But Hagelstein *was* impressed with the men's results. "Their work on glow discharge is far more successful than anyone else's," he says. "The glow discharge experiment is a beast. There are so many things that can go wrong when you are building the system. The slightest contamination shuts the thing down. At MIT we did iteration after iteration, tried everything. The effort you have to go through to get it to work is nuts. And yet the Israelis locked into a design that was first-rate, right out of the box. Either someone is a genius over there or they were just extremely lucky. When we finally got together to ask them, we said, 'How did you get this so right?' They said, 'It was luck!' Amazing."

Part of Branover and Dardik's motive in seeing Hagelstein was to ask if he would act as a consultant to the project. "Technically, I was impressed with Branover," says Hagelstein. "He seemed a down-to-earth, solid, academic type who knew an awful lot about the area he was talking about. It wasn't clear to me how Irv's ideas about SuperWaves might relate to cold fusion. True, using the SuperWave form to input the current seemed to have an enhancing effect in the glow discharge experiment, but that doesn't necessarily mean that it is tapping into something that is fundamental to nature. It could be a dynamical effect.

"My hand-waving explanation is that loading palladium is a little bit like working with plumbing that is blocked. The current drive is a little bit like having a plunger, and if you are trying to clear out pipes, you shake fast with the plunger, jiggle things loose, you clear out the pipes, and things go through better. The increase in loading, from my perspective, is like having a microscopic plunger that works. In retrospect, it seems like something clever to try. So when they asked me to consult on the project, who was I to say no?"

Hagelstein tried to persuade Branover to go public with his results at the MIT conference, but he declined, saying they weren't ready for public exposure just yet. "I said I could understand why they wouldn't want to stir up competition with their wave-form approach," says Hagelstein. "But the basic issue is that this is a scientific community, and we should be able to learn from each other. He wouldn't budge." Then, in mid-August, just a week before the conference start date, Branover changed his mind. He contacted Hagelstein to see if it was too late to be included in the program. Hagelstein said he could accept the paper because it was so important, and that he would squeeze it into a slot on Friday, the last day of the conference. As things turned out, the presentation was moved up a few days, to Wednesday, to fill a slot left open by a no-show.

The plan was that Dardik would introduce the theory of SuperWaves and Arik El-Boher would present the data. Dardik was thrilled; this was to be the first opportunity he had had to give a SuperWaves presentation in front of an audience of physicists. "Five minutes before I was due to get up and talk, Hagelstein came over to me and said, 'You can't use the word *SuperWaves*,'" recalls Dardik. "I was stunned. How could I talk about the SuperWave Principle without using the word? I felt muzzled. I decided to skip the introduction and just let Arik talk."

El-Bohar went to the dais at the front of the room and announced the title of his presentation: "Intensification of Low Energy Nuclear Reactions Using SuperWave Excitation." He described SuperWaves as waves nested within one another and reported the results of the glow discharge experiment, pointing out that excess generation almost doubled when the current was switched from direct to SuperWave form, making it many times higher than anyone else had ever achieved. "The SuperWave form was obviously having an important effect," he said. "The idea of using SuperWaves to enhancing the probability of low-energy nuclear reactions was first proposed by Dr. Irving Dardik."

"I was listening to Arik say all this stuff about SuperWaves, and I was thinking, 'Oh, my God, Hagelstein must be having a heart attack," says Dardik.

He wasn't. Hagelstein's decision to prevent Dardik from talking was tactical, and he had achieved his objective. "I know this particular community," Hagelstein told me. "If Irv had gotten up there and gone into his theoretical description the way he does, a minute and a half into it these guys would have started to leave in droves. I told Irv, 'These guys are experimentalists. Show them an experimental result first. Give them a reason to be interested, then you can get into the models if you want to.' In applied physics, that's the way things work. I know I could have done a better job handling Irv, but I really had their best interests at heart."

Dardik was definitely miffed, but that soon passed when it became clear that the presentation was a huge hit, the buzz in the hallways. On the final day, Mike McKubre presented a summary of the conference and fingered the SuperWave paper as the single most important presentation of the meeting. Martin Fleischmann was at the MIT meeting, but he missed the Israelis' presentation. Later he wrote the following to George Miley, an expert on cold fusion at the University of Illinois: "At the

Meeting, I also had a 'coffee break' during the presentation of the substitute paper emanating from Israel. I thought, right, I don't want to listen to a paper dealing with complex wave patterns—I'll go and tank up. My mistake—Mike McKubre told me afterwards that it was the best paper presented at the Meeting."

McKubre, a New Zealander by birth and by dry wit, has had more experience with electrolytic cells of the Pons and Fleischmann variety, and with cold fusion, than probably anyone on the planet. For a while, he was in the same department at the University of Southampton, England, as Fleischmann, and then moved to SRI International. He has worked on cold fusion continuously from the day it was announced, in March 1989, at the Menlo Park lab.

"The Israeli group achieved in two years what others have spent a decade trying to do," McKubre told me. "Their achievement is truly amazing, and I know of what I speak! By inputting the current in SuperWave form, they achieved loading of deuterium onto palladium at a rate and to an extent that surpasses what anyone else has achieved. And the excess heat they got surpasses what anyone else has achieved. There is nothing in my experience that would have allowed me to predict that any wave form of any sort could significantly improve the uptake of deuterium by palladium. I told Sidney Kimmel that either there really is something in Irv's SuperWave theory, or there is divine intervention here, to get the results they get."

McKubre is more prepared than Hagelstein initially was to accept that the SuperWave form in the cold fusion cells is more than just a mini-plunger effect in a micro-plumbing system. "It sounds to me as if the SuperWave Principle is *not* the babblings of a deranged mind," he says. "I can accept the SuperWave idea in medical science, but of course I don't know much about that area of science; Irv does. When it comes to physical science, I am now the expert, and I can perceive of nested waves of the sort that Irv talks about as being important in many physical systems. So my belief is that Irv knows something. How he knows it, I don't know. What *exactly* he knows, I don't know, but it does appear to be relevant to the physical world as well as the medical world."

McKubre describes Dardik's approach to cold fusion systems as "one piece of electrical engineering and one piece of magic." The electrical engineering is the construction of current input in a series of nested waves, which are fractally related to one another. Each higher-frequency wave relates to all previous wave forms. "The mathematical formulation of that, I can understand," says McKubre. The "magic" he refers to is that by manipulating waves on a macro level—the current input in a SuperWave form—you can affect the system at the molecular and atomic levels; this is the loading of deuterium onto the palladium and the promotion of fusion of pairs of deuterium atoms.

"It seems that by affecting any of the frequencies in the system, you affect them all, at all scales," McKubre explains. "It is exactly analogous to manipulating waves in the human body. Irv's cyclic exercise program grabs hold of heart rate variability, pulls it up, and that influences all the other physiological systems in the body, down to the oscillation of the biochemistry. So, I prefer to accept that the SuperWave concepts *do* have general applicability to the area of science I am familiar with. Do they have applicability to all of nature? I don't know. But I would say that Irv's model of fundamental particles is as believable and as rational as the model we currently employ. I can say that the window that Irv offers me to look through is an attractive window, because it allows me to understand things that I hadn't previously understood—for instance, the whole degree of the connectedness of things. The picture of matter as being wavelike, and pointlike only in appearance, this is an easier way for me to understand the nature of liquids and solids. But my bottom line here is, do the concepts work in systems I work with? Yes, they do, and that's enough for me."

At the end of the MIT conference, Branover asked McKubre if he would be willing to be a consultant for the Israeli project. McKubre agreed. A few days later, McKubre, Branover, Hagelstein, and a contingent from the Omer lab gathered at Dardik and Godfrey's house in New Jersey. "Mike and I took the Israelis to task on everything," remembers Hagelstein. "We needed to understand exactly what was behind what they were doing. I think they interpreted our intense questioning as our being tough, but really it was our interest and enthusiasm, which translated into a barrage of questions. That was overwhelming for them, I know. In the end, we were very impressed with their technical abilities, and everyone was on board with the project." Dardik and Kimmel now had big names from MIT and SRI International behind him, and that was a significant landmark for the SuperWave Principle.

Lesin, El-Boher, and their Israeli colleagues went back to the Omer lab and began to tweak their experiments, based on their discussions at the MIT meeting and the gathering at Dardik's house. By the following spring, they were beginning to get excess heat with the electrolytic cell more consistently than they ever had before. "We changed the wave form so that it included five nested waves on top of the carrier wave and how it was applied," explains Lesin. "We repeated the wave every 100 milliseconds. Then pause. Then reduce the current, and start over again. The pause was for the system to relax. In April, we were seeing excess heat in one experiment that continued for 340 hours. That was fantastic!"

Then, on May 18, something *really* fantastic happened. "We set up the cell and let it run overnight," explains Lesin. "When we came in the next morning, we couldn't believe our eyes. We looked at the data on the computer screen and we saw excess heat of 25 times energy input. That's more than 10 times what anyone else in the field would think is a powerful result. It lasted for 16 hours, and we were getting 25 watts out for one watt input. That's enough for you to light a 25-watt lamp using just one watt. We checked everything to make sure it wasn't a measuring artifact, and it all checked out. I called Irv right away and said, 'Irv, are you sitting down?' Then I told him what happened. He was bowled over. Everyone was." Three days later, after the cell had been revamped, a second mega-event occurred, this time producing 20 times excess heat, which lasted for 80 hours.

This was a landmark in cold fusion research, perhaps the harbinger of the first practical application, a path to sustainable energy in a world that one day is destined to run out of today's principal source of energy, fossil fuels. Whether or not this proves to be the case, we can say that the experience with the experiments in cold fusion are *consistent* with the predictions of Dardik's SuperWave Principle.

Early in 2003, Branover realized that SuperWaves might be applied to a field with which he had been familiar with for years in his work with magnetohydrodynamics, the mixing of molten metals. "In continuous casting of steel, for instance, you are working at temperatures of 1,500 Celsius," explains Branover. "That environment is very corrosive, destroys everything. In order to get strong metal, it is important to have efficient mixing. You can't use a propeller to do that because it would be destroyed in an hour. The alternative, which has been employed for decades, is electromagnetic stirring, or EMS."

EMS involves electromagnets placed around the vat containing the molten metal. Electric current is run through the magnets in a simple sine wave form, in such a way that a magnetic field effectively races round and around the vat, forcing the metal to follow it. The resulting turbulence in the metal causes the mixing action. The greater the velocity of the circulating magnetic field, the greater the velocity at which the molten metal moves, which results in greater turbulence and mixing. The downside to this is that high turbulence also leads to faster corrosion of the vat, which therefore has a shorter life; and generating high velocity consumes more energy.

"The mathematics of the process are very well understood," says Branover, "and it occurred to me that we use those equations to formulate a SuperWave power input to the electromagnets. If Irv's theory is right, that ought to produce greater turbulence more efficiently. So I had two motives here: to give some validation to what we were doing in cold fusion using SuperWaves; and to improve technology in metallurgy."

Lesin and others at the Omer lab, including Arkady Kapusta, who was brought in specifically for the project, began talking to people in the steel industry to figure out how to construct a laboratory-size vat that would faithfully reproduce the dynamics of industrial-size equipment. Kapusta directed two kinds of experiments initially. In the first, he used an alloy of indium, gallium, and tin, which is liquid at room temperature, from which he was able to determine the kind of turbulence generated by SuperWaves as compared with the traditional, simple sine wave. The second kind of experiment used an aluminum alloy heated to 760°C. When cooled after mixing with SuperWaves, microscopic analysis of the solid alloy would give an indication of the efficiency of mixing.

"The mixing experiment worked the very first time it was tried," says Branover. "We experimented with several different forms of the SuperWave to find the optimum form. The energy of the turbulence was several hundred times higher than with a sine wave, which led to more homogenous, stronger metals. That is a very important advance for the metallurgy industry." A surprising result was that not only was the turbulence induced deep into the melt, but the drag on the walls of the vessel was substantially reduced. This lessens damage to the walls and therefore increases the life of the vessel, another economic benefit of the SuperWave approach.

In February 2004, Lesin and Kapusta visited four major steel companies in the Pittsburgh area, where they presented their results. "They were uniformly impressed with what we showed them," says Lesin, "and we are now in conversation with a major equipment manufacturer for the steel industry, in Germany, to see how we can modify standard equipment to incorporate SuperWave production. But as impressed as the Pittsburgh companies were with what we were showing them in regard to better mixing, they each said they had an even bigger problem when they are casting."

The problem is this: Before molten metal is poured into a mold, it is first drained into vessel called a tundish, which is shaped like an upside-down, blunted cone. As the metal pours out of the tundish into the mold, at some point a vortex develops, just as bathwater does when going down the plug hole. Because there are always impurities on the surface of the molten metal, the tundish operator has to stop the flow before the tundish is empty, to prevent contamination of the metal that ends up in the mold. As a result, at least 2 percent of the metal in the tundish has to be held back each time, which has to be remelted, and the process started over again. Industry-wide, the cost of this inefficiency amounts to some $9 billion of lost production. "They told us that if we could reduce the amount kept in the tundish by even a small amount, a great deal of money could be saved," says Lesin.

The challenge, therefore, was to delay the point at which the vortex begins to form as the metal flows out of the tundish. "We designed a model of a tundish we could work with in the lab, which would have the same behavior of metal flow as those used in industry," says Lesin. "The first thing we tried was generating a magnetic field at the exit of the tundish, just at the sides, not all around. That made no difference to the behavior of the vortex. Next we replaced the magnetic field into a rotational one, going in the opposite direction of the vortex. We found that even a simple wave form, not the complex forms we use in cold fusion and metal mixing, significantly slowed the formation of the vortex." That work started in late summer of 2004. By early 2005, the reduction in vortex formation was up to 70 percent. "This means that the amount of metal lost each time is reduced from 2 percent to just 0.6 percent," says Lesin. "That represents an enormous potential savings to the industry. We are now looking to upscale this to industry size."

"I would say that the metallurgy experiments are working even better than we could have hoped," says Branover. "We do seem to be in position to have an impact on the efficiency of technology in metallurgy. And it gives us reason to believe that the basis for applying SuperWaves to cold fusion is correct."

The parallels between Fleischmann and Pons's story with cold fusion and Dardik's own story with the SuperWave Principle are uncanny. First the Utah chemists are hailed as visionaries (the initial response to their announcement), and then torn down by the establishment (the U.S. Department of Energy's damning report). Further work (by other scientists) vindicates them. In Dardik's case, the *New York* magazine article lifted him to guru status (which didn't especially please him, incidentally) and people flocked to benefit from his program. Then he was torn down by the decision of the New York State medical board. Further studies (his own, on healthy women, as well as Parkinson's and HIV patients, described in the previous chapter) offer a measure of vindication. A further parallel is that flying pigs were evoked in both cases!

The tests of the theory described in this chapter suggest that Dardik really is on to something fundamental, going beyond biology and medicine and into physics. As we saw, Peter Hagelstein believes that the positive outcomes in cold fusion may be merely an epiphenomenon, having nothing to do with deep theory. Mike McKubre, on the other hand, is inclined to believe that Dardik really does "know something."

12

Causing Health

All things by immortal power,
Near and Far
Hiddenly
to each other linked are,
That thou canst not stir a flower
Without troubling a star.

—FRANCIS THOMSON

THE NEOLITHIC TEMPLE OF HAGAR QIM stands majestically atop the limestone cliffs on the south coast of the tiny island of Malta, situated 100 miles south of Sicily and 300 miles east of Tunisia, in North Africa. The temple was built 5,000 years ago by a people who, to judge from archeological evidence, lived peaceful, egalitarian lives in which women played a prominent role in temple rituals, perhaps as priestesses to the Magna Mater, or Great Mother Goddess. (Malta was one of the last holdouts of goddess cultures, which had dominated human society throughout much of the Old World during the Paleolithic period, from about 35,000 years ago onward. They were replaced by patriarchal, nonegalitarian cultures.) Constructed from huge blocks of soft globigerina limestone, Hagar Qim is all curves: its megalithic outer walls form a slightly blunted oval structure when viewed from above and measure 60 meters along the long axis; its half dozen inner rooms, or apses, are all circular. The effect of this ubiquitous curvaceousness creates the paradoxical impression of a solid, immovable structure constantly in motion.

When the temple was excavated in the early 20th century, a small clay figurine was recovered from one of the inner rooms; a three-foot-tall, decorated stone altar was found in the same room, presumably a sacred place for special rituals. Headless, and with voluptuous breasts and strong thighs, the figurine came to be known, predictably, as the Venus of Malta. But it is the position of her hands that is especially significant. Her left hand rests on her stomach, while the right hand is on her thigh, pointing downward. It's a gesture common to a few other figurines at this site and at several of the other half dozen similar temples on the island. The silent but powerful gesture has been interpreted as indicating the dual, but linked, sources of life, of energy, for the people of the temples: an inner source, from the body, and an outer source, from the Earth on which they stood. Even in its stillness, it is a poignant, rhythmic gesture that seems to connect everything in life to everything else. In a sense it also evokes the ancient Hindu dictum: "the Earth is our Mother, and we are all Her children."

The people of the temple societies of Malta lived simple agrarian lives, tilling the rich, red soil and domesticating animals, some aspect of which is captured in their iconography. In another temple in Tarxien, four miles southeast of Valletta, the island's capital, domestication is depicted in a carving on a slab of limestone: a sow is shown suckling 13 offspring. The temple people, like those in hunter-gatherer societies that preceded the agricultural age, lived in close harmony with nature, the seasons, and the cycles of the moon. They did so for practical reasons, such as maximizing their harvest by knowing the best time of the year to prepare the land for planting, as well as when to reap; and for husbanding their animals at the right time of the year to ensure healthy and numerous offspring.

Like all societies living in harmony with nature in history and in prehistory, the people of Hagar Qim revered the moon, as a symbol of waxing and waning fertility and an embodiment of waxing and waning energy, an endless and infinite cycle of being. Folklore is replete with stories of people ordering their lives around the coming and going of the new moon and the full moon. "From far-northwest Greenland to the southernmost tip of Patagonia, people hail the new moon as a time for singing and praying, eating and drinking," writes historian Daniel Boorstin in his book *The Discoverers*. "Eskimos spread a feast, their sorcerers perform, they extinguish their lamps, and exchange women. African Bushmen chant a prayer: Young Moon! . . . Hail, hail, Young Moon!" In ancient Greece, according to Titus, the new moon and the full moon marked when important meetings would be held because they are "the seasons most auspicious for beginning business."

Why should poets through the ages have waxed lyrical about the moon, if it is just a stone in the sky? Aldous Huxley, in a 1931 essay, "Meditation on the Moon," had this to say: "The moon is a stone; but it is a highly numinous stone. Or, to be more precise, it is a stone about which and because of which men and women have numinous feelings. Thus, there is a soft moonlight that can give us the peace that passes understanding. There is a moonlight that inspires a kind of awe. There is a cold and austere moonlight that tells the soul of its loneliness and desperate isolation, its insignificance or its uncleanness. There is an amorous moonlight prompting to love—to love not only for an individual but sometimes even for the whole universe. . . . There are unreasoned joys, inexplicable miseries, laughters and remorses without a cause. Their sudden and fantastic alternations constitute the ordinary weather of our minds. . . . Even dogs and wolves, to judge at least by their nocturnal howlings, seem to feel in some dim bestial fashion a kind of numinous emotion about the full moon." Every one of us has the capacity within us to experience these feelings, if we are open to them.

Standing, as I did two years ago, and watching the full moon rise above the low, craggy rise behind Hagar Qim is to experience a connection with a people long gone, a way of life long gone. They were a people who experienced a connection with each other, with the earth, and with their universe in an utterly visceral way. It was part of their nature, part of their being. It was a kind of spirituality that transcends many and all dimensions, and no doubt it was explicit in the way they described their world and in the way they revered it in their songs and incantations of their pagan religion.

When Dardik set out on his journey to uncover the nature of nature, the word *spirituality* was not on his mind. In fact, as he now says, "as a vascular surgeon, I was very dismissive of ideas like that." No longer. "I now see that the SuperWave Principle does have a spiritual element to it. Not in a New Age kind of way, but as the recognition that order has emerged in the universe because of its very nature, that life has emerged, because of its very nature. For me, that does evoke a kind of awe. I'm not talking about God as a creator, but you could think about God as waves waving. Nature being what it is, it encompasses the relative order/disorder that is an absolutely perfect phenomenon at all scales, like a universal consciousness."

The spirituality that arises from the SuperWave Principle is therefore the same kind of spirituality that the people of Hagar Qim experienced, that all peoples who live in harmony with nature experience. It a spirituality of knowing that everything is connected to everything else in the universe. The route to that knowing is different, that's all: with the SuperWave principle, it arises from intellectual insight, while for the people of Hagar Qim and others, it arose from individual and collective deep experience of the world. Each of us has the innate capacity to have that same experience, just as we have the innate capacity for biophilia, or love of nature, which Harvard's E. O. Wilson has written about so eloquently. And many of us in today's industrialized societies do have a sense of universal connectedness, to nature and to our fellow humans. When we walk through the woods in the springtime, with new life budding all around; when we stroll along a mountain ridge and watch hawks hanging on the wind; when we sit by the ocean and smell the sea air and watch life in rock pools; whenever we do any of these things, do we not physically experience a oneness with nature and with our fellow beings? We do.

But for the most part, people in industrialized countries inhabit environments that stifle that connection, that cut us off from the biological nexus in which we evolved physically, psychologically, and emotionally. As a result, our individual wavenergy is torn from the supporting hierarchy of wavenergies around us; it becomes flattened, or less complex, and the normally robust oscillations of our physiology also become flattened. The outcome of all this is both dis-ease and disease. A small window into the healthfulness of our innate connection with nature comes from the simple but powerful observation of people recovering in hospitals. The presence of a single plant in the room, or a window out of which the patient can see a natural environment, trees and meadows and such, significantly reduces the time needed for recovery and speeds departure from the hospital. In industrial societies we are, as Dardik puts it, "like caged animals in a zoo, yanked from our natural, health-enhancing habitat." Many of us enjoy visiting the zoo to see exotic creatures close at hand, but most of us are unaware of the truly depressing backdrop to what we are seeing. It's true that occasionally we notice an elephant in its enclosure rocking back and forth for hours on end; or a lion, pacing back and forth, back and forth; or a gorilla eating its own feces. These, and other repetitive actions, are now recognized as abnormal, neurotic behaviors, which have been termed zoochosis.

What we don't see, however, is that the Asian elephant rocking in front of us is destined to live only 15 years or so, which compares with 30 years of hard life in a timber camp and 65 years in the wild. We don't see that most of the animals are some 50 percent heavier than they would be in the wild. We don't see that fertility is typically much lower than in the wild. And we don't see that the animals suffer high levels of chronic diseases, which are rare in the wild. "I see this as an exact parallel to the human condition in industrialized societies," says Dardik, "where chronic disease is a huge and growing crisis."

What an irony it is that chronic diseases, including heart attacks, strokes, hypertension, cancer, diabetes, multiple sclerosis, emphysema, and cirrhosis, account for the majority of deaths in all industrialized countries. We enjoy every material benefit and every medical benefit that modern society has to offer, and yet in the United States, for instance, seven of every ten deaths annually are from one or another of these so-called diseases of civilization, and 90 million people labor under a reduced quality of life because of them. These diseases consume some 75 percent of the $1.5 trillion annual health-care budget in the United States, which amounts to $4,500 for every man, woman, and child in the nation. And this grim picture continues to become yet grimmer each year, despite massive sums poured into research.

In 1971, President Nixon launched the "war on cancer" and promised to halve cancer rates in a generation. Instead, cancers in men have *increased* by 56 percent over that period, and in women, about 22 percent. Some notable exceptions aside, the incidence of most cancers continues to rise each year: breast cancer, by 0.6 percent; lung cancer in women, by 1.2 percent; melanoma in males, by 4.1 percent; prostate cancer, by 2.2 percent; colon cancer in men and women, by 2.8 percent. All this, despite many claims to the contrary by the cancer establishment, and despite $55 billion spent on research in the United States since 1971.

In January 2000, a British medical journal, *The Lancet*, carried an editorial titled "Overoptimism about Cancer." It noted that the rosy talk by researchers and charities about successes in combating cancer is designed to maintain public confidence in the establishment, so that the Niagara of government funding and public donations would not dry up. The editorial ended as follows: "Such confidence will be shattered when the public starts to see the gap between what is being said and what is being achieved." The situation with *all* chronic diseases is similarly dire, with conditions such as diabetes and cardiovascular disease reaching epidemic proportions. Every industrialized nation faces the same kind of discouraging numbers.

The tragedy, of course, is that while chronic diseases are the most common of all diseases and are among the most difficult to treat by traditional means, they are, by their nature, preventable. Studies of people in foraging societies in Africa, Australia, and South America show time after time the almost complete absence of chronic diseases of all kinds: little or no cancer; little or no cardiovascular disease; little or no diabetes; and so on. But when such people adopt a Western lifestyle, they rapidly display incidences of chronic diseases comparable to those in industrialized nations. !Kung San (Bushmen) in Africa, Eskimos, Polynesians, Native Americans, Australian Aborigines—all have gone down this path. The obvious inference is that there is something about "civilized" living that is responsible for chronic diseases.

But, some people object, isn't it also obvious that when people live longer, as they generally do in industrial societies, they inevitably become susceptible to these diseases? Not so. Young people in industrial societies show many signs of incipient chronic disease, whereas people of similar age in foraging societies do not. And people in foraging societies who live into old age continue to be free of these diseases.

What is it about civilization that causes diseases of civilization? Nutrition is a likely factor. Meat from wild animals, for instance, contains only one seventh the amount of fat, especially saturated fat, of that found in modern domesticated animals. Salt and sugar intake is much lower in foragers' diets, while fiber intake is much higher. Modern diets are therefore *discordant* with the context in which we, as a species, evolved: our digestive system and our physiology were "designed" over a period of as much as 2 million years for a certain nutritional profile, while what we put into our mouths today is significantly different from that. Does that discordance spell trouble for our health? Probably. Just how much of a factor nutrition is in causing chronic disease is hard to know, however.

But, as we have seen earlier, Dardik argues that discordance of another kind is at least as important as the kinds of food we eat, and probably significantly more so. It's a discordance of *rhythm*. For 80,000 generations, *Homo sapiens* and our ancestors, as individuals and collectively, followed the rhythm of a hunting and gathering lifestyle. People woke, slept, and pursued the food quest according to nature's daily, monthly, and seasonal cycles. And, as we saw in earlier chapters, these cycles are impressed in our DNA and the physiology and behavior it orchestrates. Four hundred generations were spent in the rhythm of life as simple agriculturalists, where, again, nature's cycles were dominant. The Industrial Age began just ten generations ago, while the press of modern living is barely two generations old. We are, in our physiology and innate rhythms, hunter-gatherers living in a world whose pace and rhythm are completely foreign to our fundamental biology. If we forget that fact, we do so at our peril.

Many industrialized nations are struggling to deliver health care to their citizens, and some face real crisis. The United States, the richest country in the world, finds itself among that latter group, where the equivalent of the Perfect Storm—complete disaster—is waiting to happen. Some of the problems are structural, such as the fact that the medical system itself is the third leading cause of death, behind cancer and heart disease, through fatal errors of various kinds in hospitals; and an insurance system that leaves 45 million people not covered. But numbers tell the real story, numbers relating to cost of the system and its efficacy. The United

States devotes 15 percent of its economy to the health-care business, some $1.5 trillion—twice as much per capita as most industrialized countries—and yet the system's performance lags behind that in most other industrialized countries. According to the World Health Organization, the United States places 41st out of 191 countries in infant mortality, behind the Faroe Islands and Guam; and 48th in life expectancy, lagging Puerto Rico and Cyprus. In overall performance, calculated from a dozen measures, the United States recently ranked 37th, compared with, for instance, the 18th place of the United Kingdom, where health-care spending per capita is less than half that in the States.

Clearly, something is extremely rotten in the state of the health-care system of the United States. An epidemic of clinical obesity definitely contributes to these numbers, but, according to some observers, so too does the increasingly heavy emphasis on specialists, compared with primary-care doctors. A quote by the turn-of-the-century Canadian physician William Osler is pertinent to this point: "It is much more important to know what kind of a person has a disease than what sort of disease a patient has." Specialists are interested in a patient only as the bearer of a specific set of symptoms to be eliminated, rather than as a person whose whole being has in some way contributed to the emergence of a specific disease.

Even Arnold Relman, onetime editor of the *New England Journal of Medicine* and usually a champion of the health-care business, sees its shortcomings. "Conventional physicians are often too interested in the disease and not interested enough in the patient," he writes. "[Physicians are] too inclined to use expensive technology and potent pharmaceuticals when simpler and more conservative approaches work as well. . . . Mainstream medicine, despite its many successes, still has only limited ability to change the course of many serious chronic diseases."

The term "health-care business" is actually a misnomer. It is, in fact, the "sickness business;" customers seek its services only when health has failed them. People in many industrialized countries, but especially in the United States, are fast becoming disenchanted with the cost of health care, as well as with the increasingly impersonal nature of a system that treats them simply as another target for yet another barrage of expensive tests and as a source of bothersome paperwork and astronomic malpractice insurance costs. It is little wonder, then, that people are increasingly turning to alternative or complementary aides to health, often tapping into simple remedies with thousands of years of history in Eastern medicine.

In the year 2000, for instance, 42 percent of Americans used some form of alternative medicine, from herbal remedies and vitamin supplements to acupuncture and energy healing. That was up from 34 percent a decade earlier, and the number continues to rise. Americans today are spending $45 billion on such remedies—still a pittance compared with mainstream medicine, but rising fast. During the year 2000, some 750 million visits were made to alternative medicine practitioners, up 70 percent from a decade earlier. Remarkably, this exceeds the number of visits paid to conventional primary care practitioners. Clearly, people more and more are turning away from the sickness business and turning toward what economist Paul Zane Pilzer calls the "wellness business."

The medical establishment—or part of it, anyway—is responding to this market pressure. Not only does the National Institutes of Health now have a well-established and well-funded National Center for Complementary and Alternative Medicine, but more than 75 universities offer courses in alternative medicine, including such prestigious institutions as Harvard, Columbia, Duke, Stanford, Georgetown, and several campuses of the University of California system. What a different atmosphere from decade ago, when the New York medical board derided Dardik for daring to think in terms other than those accepted by the establishment.

Not everyone is happy with this development, of course. For instance, Wallace Sampson, editor of *The Scientific Review of Alternative Medicine*, is quoted in a June 2000 issue of *Science* as dismissing such centers as having been "developed and driven by advocates," that the alternative medicine movement itself is "really a secular religion" that poses a threat to scientific medicine that is "more serious than anyone realizes."

"The problem here," says Dardik, "is that 'scientific medicine' is increasingly divorced from the reality of what health is. I'm not saying there are no benefits in modern medicine; of course there are. Modern medicine is very good at fixing specific problems with surgery or drugs, some of which are described as 'miracle drugs.' One such drug is Enbrel, produced by Amgen, which dramatically alleviates symptoms of rheumatoid arthritis, and thousands of arthritis suffers are very grateful for that. But this is also an example of the way the sickness business works: it fixes a problem in an individual who shows up at the doctor's office with a specific set of symptoms. The sickness business doesn't view the individual as a whole person whose symptoms are related to more than some subset of physiology that's gone awry. It doesn't ask *why* the individual got sick or *how* this relates to the hierarchy of systems, from molecular to behavioral, in his or her body. Enbrel could probably have helped me with my symptoms from ankylosing spondylitis, but I was able to eliminate those symptoms through cyclic exercise, which enhanced my heart rate variability, which in turn enhanced the health of all physiological systems in my body. It is a cascading, holistic effect."

Modern medicine is very much like modern science, in its mind-set and approach. As we saw earlier, modern science treats nature as "a well-formed puzzle" that is to be understood by taking it apart, scrutinizing each piece separately, and then expecting to understand how the whole works. This reductionist approach has had a lot of successes, but they are limited and do not give the whole picture. The SuperWave Principle does offer the whole picture; it sees nature as an infinite hierarchy of super-looping wavenergies from which order emerges at all levels; everything is interlinked, interconnected.

Similarly, modern medicine views individuals as if they are collections of subsets of physiologies that can be addressed and treated independently. Specialization is to medicine what reductionism is to science. As reductionism has had successes, so has specialization, but it is also limited, because it doesn't embrace the human body as a whole. The SuperWave Principle *does* view the body and its interlinked, interdependent physiologies as a whole, and it offers a radically different perspective on health and disease. While the sickness business *fixes disease*, the SuperWave Principle is about *causing health*, using the pervasive effects of the cyclic exercise protocol to engage the physiological systems of your body, at all levels.

"We are on the cusp of a major crisis in health," says Dardik, "and this goes beyond the practical shortcomings of the sickness business itself. The lifestyle of most adults is becoming more and more distanced from the rhythms of nature, in the places we work and even in our homes. And kids are more and more committed to sitting in front of the computer, playing video games or surfing the Internet, instead of being outside, running and playing rhythmically as kids do naturally. Look at the proportion of adults on antidepressants. Look at the proportion of kids taking antidepressants or taking Ritalin. We're doping our kids to be less depressed and down, and we are doping our kids to be less up and hyperactive. We have no idea of the long-term health effects of these drug regimens, which flatten the body's waves. And we have no idea of the long-term health effects of kids spending so much time sitting down, from an early age, which also flattens the body's waves. But from the perspective of SuperWaves, we can only expect an alarming future. Society as a whole needs to reexamine what we mean by the word *health*, and we need to reexamine how we approach health."

In Western society, health is typically understood as simply the absence of disease. And when someone gets sick, the Western medical approach is to ask what agent caused the disease, or what part of the machine went haywire. "When you come from the SuperWave perspective," says Dardik, "health is a state of maximum organization of, and communication among, all the hierarchical systems in the body, from molecular systems, to physiological systems, to organ systems, to the whole organism. And that organization and coherence comes from these systems being in healthy oscillation in relation to each other and in sync with the rhythms of nature, particularly the circadian rhythm and the lunar cycle. The SuperWave approach looks at an individual as an integrated whole, not a collection of parts, and as part of the larger environment, not separate from it. Ill health, in this perspective, happens when there is a dissociation between the internal rhythms and the external rhythms. I therefore prefer not to talk about what causes disease but instead to ask, how do you cause health?"

Causing health is about reconnecting to the rhythms of nature, reclaiming the force of natural rhythms, which is achieved by doing the cyclic exercise protocol in sync with the circadian rise and fall of energy, and the waxing and waning of the moon, as we described in chapter 10. It is not about trying to live the life of hunter-gatherers; it is about tuning your body to the rhythms that orchestrated their lives and shaped us as human beings. When we are in sync with these natural rhythms, we are at our healthiest. Western society is obsessed with "looking good." This sometimes takes the form of looking good through cosmetic surgery, for which 8.7 million Americans spent $9.4 billion dollars in 2003. It might be trying to look good by struggling with impossible diets to be anorexically thin; or by working out at the gym, hoping to get six-pack abs, a sign of "being in good shape." But there is a limited connection between looking good and being healthy. "Causing health," says Dardik, "is not about outward appearance; it is about *inward* health through shaping healthy waves."

Causing health is about taking charge of your own health, not viewing it as being the responsibility of a doctor to fix it for you with drugs and medical technology. "The biggest pharmacy is inside your own body," says Dardik, "which is actively writing prescriptions for you all the time. It is finely tuned to your needs, when you are resting, when you are active, when you are facing stress. The enzymes and hormones throughout your physiological systems are in active communication all the time, like the lily pads bobbing up and down on the pond. Causing health is about synchronizing patterns of communication throughout your body, following the rhythms of nature. When you are shaping healthy waves through cyclic exercise, your own pharmacy will be in optimal shape to help ward off disease, particularly chronic disease."

Causing health through cyclic exercise, then, is at the vanguard of the shift of emphasis from the sickness business to a more natural and rewarding wellness business. "Imagine the impact in terms of personal suffering, premature death, and the economies in industrialized nations if collectively we could embrace the holistic view of individual health that flows from the SuperWave Principle," says Dardik. "Imagine a society where people took charge of their own health, by reconnecting with the rhythms of nature, becoming in harmony with the rhythms of nature, as our ancestors were. What a triumph of human spirit that would be!"

But causing health is also about more than an individual's physical health. It is about creating powerful waves of relationship, to your partner, to your family, to your workmates, to your community, and to the environment. "Families, workplaces, and communities are healthier when they experience a palpable rhythmicity," says Dardik. "We've probably all experienced social situations that have felt flat or destructively chaotic. That happens when people are not paying attention to the positive rhythms of relationships with those around them, and the whole situation becomes dis-eased. It's a fractal of what happens when an individual is not paying attention to the rhythms of nature."

As for the natural environment, Dardik, as are many biologists, is alarmed by the destruction humans are wreaking, particularly in the rain forests and in the oceans. "We depend on a healthy environment for our *own* health," he says. "Think of the patient in the hospital, growing healthy faster, because of the presence of a plant, or a view of nature. Globally, we are like that patient, being sustained by nature. But you know, metaphorically, someone is threatening to take away the plant, to cover the window with drapes, closing off our vista on nature, dampening our healthy waves. At the same time, we are dampening the healthy waves in the environment itself. We cannot continue to go down that path and expect to remain healthy as individuals and as a species."

13

What You Can Do

"Humankind has not woven the web of life. We are but one thread within it. Whatever we do to the web, we do to ourselves. All things are bound together. All things connect."

—CHIEF SEATTLE

SO FAR, THE INDIVIDUALS I'VE DESCRIBED who have followed the cyclic exercise protocol have been victims of some kind of chronic disease, and you've seen dramatic improvements in many cases. Here's an example of a healthy person's experience on the program—Stephanie, one of the participants in the healthy women's trial we described in chapter 10.

Stephanie is director of patient safety and risk management at Hunterdon Medical Center in New Jersey. The oldest of eight siblings, she learned early on the power of her own intuition, as she acted as a third parent to three brothers and four sisters. Stephanie still brings powerful intuition to her work, as well as energy, passion, and relentlessness, according to a close colleague. Her choice of a nursing career was very pragmatic, a "job security thing," as she puts it. Pushed by her parents, Stephanie was following "the script" for what a young woman was expected to do in terms of career in the 1970s. Linguistics, or rather the psychology of language, was her passion, but her parents that deemed as being an "airheaded" thing to do.

As the person charged with getting bureaucracies to change their thinking and behavior, to improve patient care and reduce risk, she hears true feelings encased in people's language. "Instead of admitting they don't *like* the new ideas," she says, "people will say, 'We've always done it this way, and it works'—even when it doesn't work—or, 'It will take so long to do.'" She now finds herself finely tuned to the words people use in her current work endeavors, saying, "Their choice of language gives me a window into what they are really thinking." Her early interest in linguistics is at the fore again. "I guess I've come full circle," she says.

In 1999, Stephanie joined 10 of her nursing colleagues on a clinical trial of Dardik's cyclic exercise protocol, run by researchers at Harvard and Columbia universities. The study was designed to measure the impact of the protocol on various physiological and psychological parameters in healthy individuals. Here she describes her experience of the protocol:

"I heard Irv Dardik talk about his cyclic exercise program at a conference on complexity, in New Jersey, sometime in 1998. He was talking about how everything is connected to everything else in the body, and about the importance of reconnecting with the waves and rhythms in nature. I was giddy with joy, don't really know why. But I felt such a profound resonance with what he was saying. It just felt right.

"About a year after that, I was asked if I wanted to be part of the clinical trial, and I jumped at the chance. I was working out at home at the time, with weights and aerobics. But I wasn't in the best of health. I was 60 pounds overweight; I was energetic—you have to be in my job—but I wasn't sleeping very well. I would get sick three or four times a year, and my back would go out a lot. I thought, 'Why not give it a go, see what happens.'

"Donna Cole, who was coordinating the program, set up a room in the new Endoscopy and Pain Center at the hospital. She had all the exercise equipment there, and the recording equipment. We would do exercises at different times of the day, depending on where we were in the lunar month. I like that connection, with the lunar cycle.

"At first I used the trampoline, holding weights, to bring the heart rate up, just for a minute, and then rest to bring it back down again. And so on, maybe three or four times. We were encouraged to meditate during this recovery period, to help the recovery process. I'd already been doing some meditation before the trial, so that was easy for me. Pretty soon I found that the trampoline exercise wasn't spiking my heart rate enough, so I started using a NordicTrack at home. That worked, and it made it easier to stick to the program.

"After about a month, I really began to feel a significant difference. I felt more sustained energy during the day than I used to. I am a pretty focused person anyway, but I definitely handled stress better; I wasn't so all over the place emotionally, as I had been. I no longer had the high/low mood swings, could handle the intensity of my role with greater balance; in risk management, you are always putting out fires of one sort or another. I slept through the night, not waking at two or three in the morning as I used to. It was great. And my period aligned to the lunar phase, the new moon, which is most appropriate for inward quiet. That's the way it's supposed to be.

"The most surprising thing to me was sleeping. I found myself sleeping better and waking up refreshed in the morning. And for the first time in my life, I started remembering my dreams clearly; I started writing them down in the morning. The dreams were phenomenal, very vivid, and they seemed very healing. I don't know if it was a coincidence, but all of a sudden there was a flow in my life, in my personal life, in my work, in my well-being. It was amazing. In the year after the trial, I kept on doing the cycles; I didn't get sick, and my back didn't go out like used to. And I subsequently lost those 60 pounds!

"I have to say that this period was one of the most transformative times in my life, connecting me with rhythms and cycles of nature, connecting with myself in my life. It changed how I look at and regard almost everything. I was so excited about this knowledge of fractals and cycles within it that I shared my experience with anyone who would listen. My husband's cousin had lupus, and I gave her some of the literature on the exercise program. She started exercising more cyclically, and she now has two children, no problem. The condition hasn't gone away, but she now has a lower incidence of exacerbation. My husband has steered toward cyclic patterns of exercise.

"I talked to people up and down my street about becoming attuned to lunar cycles, the natural cycles of nature, thinking about better diets; what's right for each time of year. Most of them were enthusiastic but also a little puzzled. They'd say, 'It sounds great, but how can this little exercise program do so much?' It's a good question, but I know it works for me, and I could see the same thing happening with the other people on the study.

"I've just turned 50, and I've never felt better!"

If you want to do the program as Stephanie did, you will find the way at the following Web site: www.LifeWaves.com. But what if you are like Stephanie's friends and relations, someone intrigued with the health and well-being benefits of the SuperWave Principle but, for whatever reason, can't do the exercise protocol in a formal way? What can you do? A great deal, actually—a lot of simple, practical things that can have a powerful impact on how you *feel* in terms of health, how you *are* in psychological well-being, how you *perform* in your work, and how *creative* you can be. Every aspect of your life is touched: self, family, community, work life. Some of these actions can seem somewhat spiritual, behaviors that too often in our society are derided as "soft," "touchy-feely," or "New Agey." In fact, these behaviors reinforce and enhance the benefits that flow from the practical, down-to-earth actions that are guided by an awareness of waves in your life. Continue to deride them if you must, but know that in doing so you are getting less in your life, and achieving less in your work, than you might otherwise. Godfrey, as coarchitect of the formal program, has given a great deal of thought as to how you can incorporate the LifeWaves philosophy into your everyday life. The following is based on her accumulated wisdom.

There are two aspects to the practical, everyday things you can do. The first is to make sure that you *do what you do every day in harmony with nature's rhythms*: the lunar cycle; the daily, or circadian, cycle; and cycles shorter than a day, the so-called ultradian cycles. This will reconnect you with nature's rhythms and bring harmony into your life. The second starts with a simple premise: Humans, in our deepest being, are creatures of rhythm in a universe of rhythms. Waves are everywhere in us and in nature, as we saw in earlier chapters. Health in ourselves and in nature depends profoundly on constant oscillation. From this simple premise, we proceed to a simple road map for health, well-being, and creativity: *Notice when constancy is happening*, whether it is at a high plateau or a low trough, and *change it*. Do something that relaxes you off the high plateau or boosts you out of the trough. In other words, be on the alert to *make waves*. Here's how.

DO WHAT YOU DO EVERY DAY IN HARMONY WITH NATURE'S RHYTHMS

The circadian rhythm is the most powerful wave in all our lives. It's why we feel drowsy in the evening and go to sleep; it's why we stir and wake up as day breaks. As we saw earlier, myriad physiological systems fluctuate in train with the circadian cycle, and these are the biochemical backdrop to a wave of energy and alertness we experience through the day. The trough of energy is in the early hours of the morning, between 3:00 and 6:00 A.M.; the energy wave begins to rise just before we wake, continuing to a peak in the mid- to late afternoon, after which it goes into decline to complete the cycle. This mid-afternoon peak is when most world records are set in athletics, and it is the time we are most alert mentally—or should be. You might well say, "Wait a minute, I know I am not alone in feeling a real slump of energy in mid-afternoon; I'm not alert at all." That's true for a lot of people, and there's a reason for it, as we'll come to see.

Pulling out of the trough in early morning is difficult for many people. Instead of heading for the kitchen for a belt of caffeine to jump-start your day chemically, do it physiologically. Take a brisk 10-minute walk around a few blocks; swing your arms vigorously; get your heart rate up; and make sure to look up to the bright sky (if the season is right) to prompt the light-sensitive cells in your eyes to set the day's rising physiology in train. When you get home, sit down, thus creating a wave of exertion and recovery that will cascade throughout your body's physiological systems. Have a protein-rich breakfast, the best way to fuel the mental processes you will need as the day goes on, instead of the typical carbohydrate jolt of a bagel or muffin.

What if you can't take the walk for good practical reasons, such as having to get the kids ready for school? A good alternative is to use your shower time to make waves, by alternating between 30 seconds of hot water and 30 seconds of cold, four or five times. Many people's reaction to this advice is to protest, "Wait a minute! I can't bear searingly hot water, and I can't take freezing cold!" It's a natural reaction; with our society's "no pain, no gain" obsession, we often think we have to push everything we do to the limit. You don't have to. Just run the water as hot as is comfortable for you and as cold as is bearable, without discomfort. You will be making temperature waves that will translate to physiological waves. Even if you do have time for the brisk-walk routine, starting with temperature waves as well is a good idea. Hot/cold showers are especially effective in preventing colds.

During your workday, make sure to take a proper break for a lunch that is, again, rich in protein. After lunch your energy wave takes a dip, and you should respect your body's need for a siesta. If you are at home, you have the luxury of taking a 20-minute nap in subdued light; you'll wake up refreshed and on the energy upswing again. At work that may not be possible, of course. But find a way of resting with your eyes closed, maybe with your head on your arms at your desk. Your colleagues might think it a little weird that you appear to be slacking off. That's a product of our push-push work culture. But you can be sure that, come mid-afternoon, you will be at your productive, creative peak, while they will be in an energy slump because they tried to override the natural circadian rhythm. Some companies are becoming enlightened about the postlunch energy dip and allow or even encourage people to take a power nap. Odd that it should be called a "power" nap when really it is a "natural" pause, but that's the macho nature of our work culture. If you want vibrant, productive office meetings, schedule them for mid- to late afternoon, when people are at the peak of their energy and cognition—assuming they haven't pushed through and ignored the earlier call to siesta, of course.

Back at home in the early evening, your energy wave is beginning to subside, and you should slow down, too, in harmony with your natural, internal rhythm. Relax, have a glass of wine, a light dinner, and early to bed (around 10:00 P.M.), in as complete darkness as you can manage. My mother used to say to me, "Early to bed, early to rise, makes a man healthy, wealthy, and wise!" She didn't know about the science of chronobiology and circadian rhythms, or the SuperWave Principle, of course, but the old saw she was repeating to me *is* based in science, though the "wealthy" part cannot be guaranteed! Studies in chronobiology show consistently that when people work in harmony with their circadian rhythm, as I describe, their mental acuity and physical energy is much improved over those who push themselves through the day and ignore the rhythm.

If you find yourself in charge of shaping a business retreat, or any kind of retreat where you need to tap into the innovative resources of a group, your chances of success will be much greater if you pay attention to circadian rhythms. Rise early and begin with some form of collective physical activity; maybe a brief burst of jogging, running on the spot, yoga, whatever you choose, as long as you are making exertion/recovery waves. People will begin to feel physically good and the group will begin to cohere as a team, having made waves together. Eat breakfast. Then incorporate as much rhythmic physical activity into learning sessions as you can, with people interacting directly with one another. Have them do mirroring exercises or tell personal stories. This will reinforce the waves of relationship that started to form in the early-morning collective exercise. Take lunch, then a nap, and continue in the same vein through the afternoon. An early dinner in subdued lighting, perhaps by candlelight, brings the group smoothly down the energy wave toward bedtime. Sit around a campfire, tell stories, read poetry, reflect on the day in a fee-flowing way, and amazing insights will emerge.

Many companies hold retreats in natural settings, and there is a reason this is indeed best: we are creatures of nature, and when we connect with nature we are better able to connect to ourselves; we are better able to connect to our inner feelings and our often untapped cognitive resources. Sounds New Agey, doesn't it? But how many of us have experienced such gatherings, have witnessed and been part of the intergroup bonding that emerges, and have thrilled at the accomplishments that can be had in such a setting? Too often, people then go back to the workplace and slip back into the unnatural environment of the office and into the unnatural roles we play there. We can't be in nature all the time, of course, but staying in sync with the rhythms of nature will keep that flame alive.

Just as music and storytelling have cycles of tension and resolution, scientists have found that our bodies' physiological and cognitive systems go through cycles of activity and recovery during the day. Riding on top of the 24-hour circadian rhythm are shorter rhythms of around 90 to 120 minutes, called **ultradian rhythms.** It's waves waving within waves, just as the SuperWave Principle describes. Google the term "ultradian rhythm," and you will find 6,300 sites describing the cyclic rise and fall of all kinds of physiological factors, such as enzymes and hormones, and of sleep patterns. Again, this is the biochemical background of a rhythmic rise and fall of physical energy and mental acuity and attention through the day.

I remember a lecturer once saying that "the mind can absorb only what the seat can endure," meaning that you can't sit for too long and still hope to focus on what is being said. In fact, it isn't your butt that gets tired; it is your mind. Just as our muscle fibers are designed for bursts of exertion rather than sustained exercise, so, too, is our mind designed to be active in bursts rather than grinding on and on. That's the ultradian rhythm, where you find your attention drifting after pursuing a mental task for an hour and a half or so. You start to yawn, gaze out of the window, daydream . . . . The "flow" starts to break. This is your body telling you that you are going into a rest cycle of the ultradian rhythm, and scientists have shown that when you ignore it and continue to try to work, the chances for errors begin to rise exponentially.

What can you do to be in sync with this rhythm? Many people find themselves responding to the ultradian rest signal by getting a cup of coffee or going outside for a cigarette. Leaving your desk is definitely a good idea, but instead of seeking a caffeine belt to restore your energy, follow the SuperWave Principle. Get your heart rate up quickly by running up a flight of stairs two or three times, or running rapidly on the spot, and then sit down and rest, nap, or meditate for 5, 10, 15 minutes. That way you will allow your body to restore itself by *going through the trough* rather than *trying to fight against it*. You will get back on the energy upswing in a natural way. Studies show that when people are allowed do this without being condemned as being weird, lazy, or slacking off, productivity increases, innovation increases, and the rate of sickness drops.

The weakest of the energy cycles we experience is the **lunar rhythm**. Women are very aware of their menstrual cycle, of course, and when they are fully in tune with nature, they will ovulate at the full moon, the highest-energy point of the lunar rhythm, and have their period at the low-energy point, the new moon. Both men and women respond to this cycle of energy, often in subtle, or not so subtle, ways. If you talk to anyone who has worked in an emergency room, you will hear that things get a little crazy around the full moon, when hospital admissions from bizarre accidents go up.

How should you be in sync with the lunar cycle in your life and in your work? If you want to run a highly energetic company retreat, where wild and unexpected things might happen, hold it around the full moon. Similarly, if you want a team project to peak at high energy, focus it around the full moon. If it's quiet reflection you seek, the new moon is the time to gather, but beware; people can become very soulful, more emotionally vulnerable, both of which can be valuable but need to be handled carefully. Again, this sounds dangerously New Agey, but studies show that people's moods do cycle with the moon's rhythms. Aldous Huxley's "Meditation on the Moon" was based on reality, not poetic fantasy.

MAKE WAVES

In Business: Why did I say "dangerously" New Agey? Because the kind of gladiator culture that reigns in many, if not most, big corporations these days boasts its independence from nature, its independence from those rhythms and our connectedness to them. It prides itself on being driven to achieve goals set entirely outside of nature, which is largely true, and then tries to do that, also outside of nature.To admit that a corporation might operate in a milieu not entirely under the control of the CEO, or that individuals might have needs that are simply not addressed in business school, is regarded as being soft and unbusinesslike. Why this has come to be is the subject of another book, not this one. But this attitude is dangerously shortsighted because corporations are like organisms themselves, with rhythms of their own, composed of people, with rhythms of *their* own. When people are released from the emotional prison that the macho business culture has erected, their talent is unleashed, and the corporation thrives.

It's amazing that the term "24/7" has become so iconic of modern society: drive yourself at a constant peak of performance through 12-hour (or longer) days; forget real vacations; be a success! It's a culture of chronic stress, anxiety, and fear. Too often, the outcome for individuals is burnout, and, perversely, that is sometimes regarded as a badge of honor. Remember the comparison of heart rate maxima in cyclic versus sustained exercise? If you run for an hour, you can sustain a heart rate of, say, 120 beats a minute. But when you do cycles, you can reach repeated peaks far higher than that—160, 170, 180—recovering between each peak. When you do this, your physiology, your immune system, is healthier.

It's the same in business. If you have an advertising pitch that is going to take a month to develop, or a new piece of software to be written by a team—any goal that has a time frame of more than a few days—don't try to go all out until it's finished. Modulate the effort in cycles of a few days of intense activity interspersed with periods of recovery. You will find that the overall productivity and the flow of brilliant ideas will be greater. And people won't burn out. From the perspective of the SuperWave Principle, you have the large wave of a few days of push; on top of that is the circadian energy wave; and on top of that are the ultradian energy waves. It is waves waving within waves, the sign of a healthy organism (the corporation). The corporation that flatlines at a suboptimal level of performance is pathological, just as flatlining in the heart is pathological.

Some business observers actually recognize this, people such as Jim Collins, author of the blockbuster best-selling book *Good to Great*; Arie de Geus, author of *The Living Company*; and Jeffrey Pfeffer, author of *The Human Equation*. Each of them identifies the most successful companies as those in which leaders foster a culture that recognizes and honors the fact that their people are not cogs in a machine that can be driven at a constant hum. Instead they know that their people, and their corporations, are living beings whose rhythms need daily, weekly, monthly expression. These leaders and their companies are not the ones typically lauded in business magazines, but they are the most successful leaders and companies in the economy. (Again, you could ask, If success is what business leaders want, why do they continue to pursue the gladiator mode rather than the rhythmic organism mode? But that's another story.)

Here's a story of a business leader who does know about the SuperWave Principle and uses it to good effect. Stephanie's boss, Linda Rush, is the vice president of nursing at Hunterdon Medical Center in New Jersey, which has a very nurturing work culture, has good financial numbers (rare in the health-care business these days), and is rated top in the state for patient satisfaction. Even at Hunterdon, however, there are times of stress, when people feel overwhelmed. On one such occasion, Linda sent out an e-mail to her nurse leaders, asking them to come immediately to her office. A dozen nurses turned up, anxious, wondering what bad news was about to be announced that would make a difficult time even worse. Instead, Linda asked people to push the conference table and chairs to the side of the room. She then turned on the CD player, from which blasted raunchy music. "Dance!" she commanded. The place was a riot, with the nurses dancing furiously, singing and laughing, having fun for 10 minutes. This simple act broke the dark mood, and people went back to their work ready to face any obstacle.

Dancing creates powerful waves of connection, and the nurses felt that viscerally; it transformed them on the spot. Not every business leader would be comfortable asking their colleagues to throw off their inhibitions and dance, as Linda did. But there are many ways to make waves to break out of a slump. Your job is to think of them—and do them.

In Your Daily Life: You can make waves in most everything you do. You need to visit a colleague two floors up? Take the stairs, not the elevator, and walk briskly, or even run up, to elevate your heart rate, and then sit down for a few minutes. You are coming out of the subway on a flight of stairs? Use them, really use them. Walk briskly or run up, and then stop, until your heart rate recovers. Don't ride up escalators—walk or run, and then stop and recover. When you are hiking in the countryside, make use of hills: climb them briskly, until you feel your heart pounding, and then stop and admire the view for a few minutes, until your heart rate is back near normal. Don't sit at your computer or in front of the television for hours; get up every so often and move around. Just standing up raises your heart rate 10 to 15 beats a minute. If you are preparing to go into an important meeting, do some cycles of exertion and recovery immediately beforehand. Your mental acuity will be sharpened, and you will perform better. Once you start thinking in this mode, it is easy to incorporate making waves into your life in all kinds of ways.

Many people exercise regularly for pleasure or fitness. People run, lift weights, cycle, play tennis, all manner of activities. The SuperWave Principle does not say, Stop doing that! It says, From time to time incorporate some wave patterns into what you love to do. Don't exercise like a lab rat running on a treadmill; do it like a cheetah runs, in energetic bursts. If you enjoy bicycling, for example, push yourself hard for a minute or two, and then coast. Do that half a dozen times. If you're a runner, occasionally run in cycles: sprint as fast as you can for a minute, and then stop or walk until your heart is no longer pounding. Do that four or five times. When playing tennis, make sure to stop a little longer between points, allowing your heart rate to come down more than it would otherwise. Tennis lends itself very well to making waves of exertion and recovery, as does weight lifting. The SuperWave Principle simply says: Consciously add occasional cycles to whatever it is you like doing for exercise. You become your own personal trainer.

You can use waves to break moods of either extreme. If you are angry or stressed, for instance, don't try to calm yourself down emotionally. Do it physically and physiologically. Stop where you are and do a few cycles of exertion and recovery, by whatever means you can. Run vigorously on the spot, run up a flight of stairs, whatever is most convenient, and then stop. In anger or stress, your physiology is stuck at a plateau. Cycling will help bring it down through a mind/body connection. If you are feeling blue or even depressed, make waves, and your physiology will lift out of the trough. It's that mind/body connection again. If your child is throwing a tantrum, get him to run around—in fact, run with him—then stop. The tantrum will dissipate. You'll feel calmer, too.

With People: A few years back, researchers at the University of Miami Medical School did a study on premature babies in incubators. They found that babies who were regularly stroked gained weight 47 percent faster than those who weren't touched. It wasn't that the stroked babies ate more. They didn't. The effect of contact with another human being was to organize their physiology, enhancing communication between physiological systems, to make more efficient use of the nutrients they took in. Not only that, at age eight months, these babies performed better on mental and physical tests than those who weren't stroked. This little story illustrates the importance to health of physical human contact, especially in children, but it surely translates to adults, too. People who live alone often become depressed, and they die earlier than those who share their life with a partner. Studies show that people who live alone can greatly alleviate their loneliness if they have a pet, particularly one they can touch, such as a dog or a cat. Apparently, the opportunity to make waves with another living creature, even if nonhuman, helps promote health.

Who was it that decided that babies should be put in their cribs at night and left to cry themselves to sleep, and why? Humans are quintessential social creatures; we thrive through contact and interaction with other humans, and suffer when that is absent. If you are a mother who prefers to have your infant in bed with you occasionally, but feel guilty about it, don't. You are making waves of relationship with your child through physical contact, forming bonds that enrich the relationship. You are resonating with each other, building a physiological memory of a healthy relationship. Continue that physical and emotional bonding, and your relationship will remain healthy.

What of the whole family? There's that old saying, "The family that plays together, stays together." It sounds trite, but from the perspective of the SuperWave Principle, you are constantly making waves of relationship and connection, leading to a healthier family group. Children grow up, of course, and become teenagers who rebel, stay up late, sleep till all hours, and generally want to lead their own lives, in which parents too often are a source of embarrassment rather than of companionship. Sad to say, the SuperWave Principle hasn't figured a way around that one yet—except to say that bonds formed early through healthy waves will remain strong.

Most of us have experienced the exhilaration and joy of bonding with a partner in a new, intimate relationship. You are fully focused on each other, fully in sync physically and emotionally, and the feeling of that is absolutely visceral. That feeling comes from powerful waves of physiological oscillation. This initial intensity isn't usually sustained, or even sustainable, of course. But the couple that can maintain a high level of paying attention to each other's needs, continuing to oscillate the waves of personal interaction, will have a strong, lasting relationship.

What about professional relationships? There isn't the intensity or the intimacy, of course, but the prescription is the same: interact as human beings and pay attention to the other's needs, rather than seeking ways to feed your own ego, or, worse, trying always to be one up on the other person.

A few years back, I conducted a study with a colleague on the factors that create the most adaptable, creative, successful work environments, which was published in a book, *Weaving Complexity and Business: Engaging the Soul at Work*. We found that the most successful work cultures were those that encouraged people to treat their colleagues as *people*, rather than as cogs in the business machine. People who work this way are creating waves of relationship, from which respect and trust emerge, which allows people to explore crazy ideas without fear of embarrassment. That's the fount of creativity, and it shows up in the organization's bottom line. The traditional business model is very mechanistic, with people viewed as automatons from which leaders try to extract maximum efficiency. The reason this perspective is limited and produces inferior results is that automatons, like machines, can't make waves of relationship; nothing powerful emerges, and creativity is dampened.

Having fun in a group is a powerful way to make waves of connection, as we saw in Linda's story earlier. Laughter is strong medicine, and

Norman Cousins credits laughter therapy for reversing his ankylosing spondylitis, which he wrote about in his book, *Anatomy of an Illness*. When people laugh together, there is a very visceral connection, which drives powerful physiological, health-promoting oscillations. Saranne Rothberg, who in 1999 founded the ComedyCures Foundation, which uses humor of different kinds to help heal sick people and sick organizations and to bring relief and psychological balance to stressed troops, has a record of great success. Early in 2004, she met Dardik and heard about the SuperWave Principle. "It was clear from the first moment that we spoke that ComedyCures is the human application of your scientific discoveries," she told Dardik. When you hear laughter in the workplace, you know you have a healthy organization. Promote that, encourage it, be part of it, rather than frown upon it as being "unbusinesslike," as too often happens.

With Nature: Gordon Orions, an ecologist at the University of Washington, Seattle, did a study some years back in which he showed his students slides of different kinds of landscape—forest, plains, mountains, and so on—and asked them which one they most strongly resonated with in a fundamental way. No matter where they grew up as kids, no matter what kind of landscape they were most familiar with, they overwhelmingly chose gently rolling countryside, dotted here and there with clusters of trees. It was the image of the African savannah, the cradle of humankind. Orions suggested that each of us has a genetic memory of the landscape of our origins, a cogent connection with nature. Remember the health benefits to a recovering patient of a glimpse of nature out a window or simply having potted plants around. That's a reflection of the deep connection with nature that Orions identified.

Each of us should follow that same prescription and be with nature as much as we can, really *be* with nature. How is it that people who do this, who really honor nature as a cogent force in their lives, are in sync with nature, are often derided as "tree huggers" or some such derogatory term? It's that gladiator, "I don't need anyone or anything," mentality. Forget that. Connect with nature, whether you take a trip to the countryside, go into a city park, or deck your house with plants, knowing that you are connecting with something fundamental to *your* nature; knowing that you are making waves that are fundamental to your health. Hug a tree if you want to, or just sit quietly in resonance with the natural environment. Be proud of that. Honor it. Find a way to really be in sync with nature, and you will find that you will be more in sync with yourself and with others around you. It's your natural state, and the very nature of nature, and it leads you down the path to enhanced health. The SuperWave Principle explains why.

Index

A

B

D

E

F

G

H

I

J

K

L

M

N

O

P

Q

R

S

T

U

V

W

Y

Z